Pipefitting
Level Four

Trainee Guide
Third Edition

PEARSON

Upper Saddle River,
New Jersey
Columbus, Ohio

NCCER

President: Don Whyte
Director of Curriculum Revision and Development: Daniele Stacey
Pipefitting Project Manager: Tania Domenech
Production Manager: Tim Davis
Quality Assurance Coordinator: Debie Ness
Editors: Rob Richardson and Matt Tischler
Desktop Publishing Coordinator: James McKay

NCCER would like to acknowledge the contract service provider for this curriculum: Topaz Publications, Liverpool, New York.

This information is general in nature and intended for training purposes only. Actual performance of activities described in this manual requires compliance with all applicable operating, service, maintenance, and safety procedures under the direction of qualified personnel. References in this manual to patented or proprietary devices do not constitute a recommendation of their use.

Copyright © 1993, 1998, 2007 by NCCER, Alachua, FL 32615, and published by Pearson Education, Inc., Upper Saddle River, NJ 07458. All rights reserved. Printed in the United States of America. This publication is protected by Copyright and permission should be obtained from the NCCER prior to any prohibited reproduction, storage in a retrieval system, or transmission in any form or by any means, electronic, mechanical, photocopying, recording, or likewise. For information regarding permission(s), write to: NCCER Product Development, 13614 Progress Blvd., Alachua, FL 32615.

13 14 15 16 17

ISBN 10: 0-13-614429-2
ISBN 13: 978-0-13-614429-8

PREFACE

TO THE TRAINEE

There are some who may consider pipefitting synonymous with plumbing, but these are really two very distinct trades. Plumbers install and repair the water, waste disposal, drainage, and gas systems in homes and in commercial and industrial buildings. Pipefitters, on the other hand, install and repair both high- and low-pressure pipe systems used in manufacturing, in the generation of electricity, and in the heating and cooling of buildings.

If you're trying to imagine a setting involving pipefitters, think of large power plants that create and distribute energy throughout the nation; think of manufacturing plants, chemical plants, and piping systems that carry all kinds of liquids, gaseous, and solid materials.

If you're trying to imagine a job in pipefitting, picture yourself in a job that won't go away for a long time. As the US government reports, the demand for skilled pipefitters continues to outpace the supply of workers trained in this craft. And high demand typically means higher pay, making pipefitters among the highest-paid construction workers in the nation.

While pipefitters and plumbers perform different tasks, the aptitudes involved in these crafts are comparable. Attention to detail, spatial and mechanical abilities, and the ability to work efficiently with the tools of their trade are key. If you think you might have what it takes to work in this high-demand occupation, contact your local NCCER Training Sponsor to see if they offer a training program in this craft, or contact your local union or non-union training programs. You might make the perfect "fit".

We wish you success as you embark on your fourth year of training in the pipefitting craft and hope that you'll continue your training beyond this textbook. There are more than a half-million people employed in this work in the United States, and as most of them can tell you, there are many opportunities awaiting those with the skills and desire to move forward in the construction industry.

We invite you to visit the NCCER website at **www.nccer.org** for the latest releases, training information, *Cornerstone* magazine, and much more. You can also reference the Pearson product catalog online at **www.crafttraining.com**. Your feedback is welcome. You may email your comments to **curriculum@nccer.org** or send general comments and inquiries to **info@nccer.org**.

NCCER STANDARDIZED CURRICULA

NCCER is a not-for-profit 501(c)(3) education foundation established in 1995 by the world's largest and most progressive construction companies and national construction associations. It was founded to address the severe workforce shortage facing the industry and to develop a standardized training process and curricula. Today, NCCER is supported by hundreds of leading construction and maintenance companies, manufacturers, and national associations. The NCCER Standardized Curricula was developed by NCCER in partnership with Pearson, the world's largest educational publisher.

Some features of NCCER's Standardized Curricula are as follows:

- An industry-proven record of success
- Curricula developed by the industry for the industry
- National standardization, providing portability of learned job skills and educational credits
- Compliance with the Office of Apprenticeship requirements for related classroom training (*CFR 29:29*)
- Well-illustrated, up-to-date, and practical information

NCCER also maintains a Registry that provides transcripts, certificates, and wallet cards to individuals who have successfully completed a level of training within a craft in NCCER's Curricula. *Training programs must be delivered by an NCCER Accredited Training Sponsor in order to receive these credentials.*

Contents

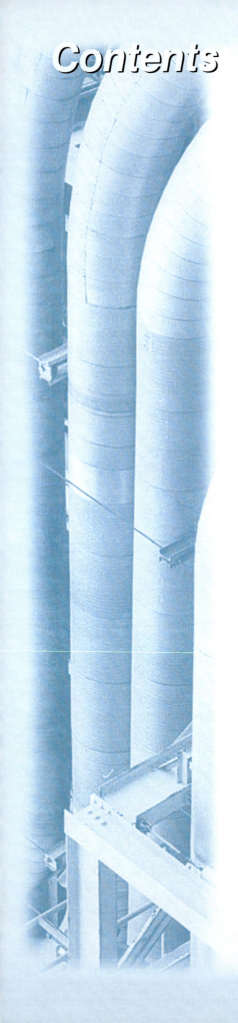

08401-07 Advanced Blueprint Reading   1.i

This module prepares the pipefitting trainee to work with blueprints used in the shop and in the field. It presents the skills needed to derive necessary construction information from P&IDs, general arrangement drawings, ISOs, and spool sheets. As a practical application, trainees are given step-by-step instructions for following a line of pipe through a set of drawings. Includes nine 11"×17" blueprints. **(50 Hours)**

08402-07 Advanced Pipe Fabrication   2.i

Discusses how to lay out and fabricate mitered bends, laterals, wyes, and ninety-degree intersections using tables of ordinates or a calculator. This knowledge is required in order to fabricate specialty bends and intersections. **(50 Hours)**

08403-07 Stress Relieving and Aligning   3.i

When piping is assembled, stresses from welds can result in misalignment and pipe strain, which can be compounded at the point of attachment to the pumps. The purpose of this module is to teach the nature of misalignment and methods of correcting it. Includes terminology that will help pipefitters communicate with millwrights who perform pump setup. **(15 hours)**

08404-07 Steam Traps   4.i

In steam systems, condensate can damage valves and fittings and reduce the efficiency of the system. Steam traps release the condensate, either to the atmosphere or by returning it to reheaters. This module describes types of steam traps, how they function, and the basic methods for troubleshooting them. **(15 Hours)**

08405-07 In-Line Specialties   5.i

Describes the various specialty devices that are used in pipelines, including: bleed rings; ball and expansion joints; measuring devices for temperature, level, flow rate, and pressure; steam traps; drip legs; and desuperheaters. The purpose and function of each type is explained. **(10 Hours)**

08406-07 Special Piping   6.i

Discusses methods of assembling copper and plastic pipe and tubing. Introduces brazing and soldering, and explains the differences between these two procedures. Also describes compression and flared fittings, and grooved and compression formed joining methods. **(25 Hours)**

08407-07 Hot Taps 7.i

Hot tapping is a common requirement, although rarely performed by most pipefitters. Trainees learn how to prepare for the work to be done by the hot tap specialist. The mechanics of attaching fittings to the pipeline while the line is under pressure is explained. Covers line stopping, freeze stopping, and adding connections to the line. **(10 Hours)**

08408-07 Maintaining Valves 8.i

Explains how to replace packing and O-rings, and how to open and close a valve's bonnet. Discusses how to safely troubleshoot and maintain several types of valves. **(10 Hours)**

08409-07 Introduction to Supervisory Roles 9.i

Offers basic information for pipefitters who have a desire to move into supervisory roles. Provides information on issues related to cultural differences, gender-based social behaviors, and legal and ethical situations that a supervisor is likely to encounter. **(7.5 Hours)**

Glossary of Trade Terms G.1
Index I.1

NCCER Standardized Curricula

NCCER's training programs comprise more than 80 construction, maintenance, pipeline, and utility areas and include skills assessments, safety training, and management education.

Boilermaking
Cabinetmaking
Carpentry
Concrete Finishing
Construction Craft Laborer
Construction Technology
Core Curriculum:
   Introductory Craft Skills
Drywall
Electrical
Electronic Systems Technician
Heating, Ventilating, and
   Air Conditioning
Heavy Equipment Operations
Highway/Heavy Construction
Hydroblasting
Industrial Coating and Lining
   Application Specialist
Industrial Maintenance Electrical
   and Instrumentation Technician
Industrial Maintenance
   Mechanic
Instrumentation
Insulating
Ironworking
Masonry
Millwright
Mobile Crane Operations
Painting
Painting, Industrial
Pipefitting
Pipelayer
Plumbing
Reinforcing Ironwork
Rigging
Scaffolding
Sheet Metal
Signal Person
Site Layout
Sprinkler Fitting
Tower Crane Operator
Welding

Maritime

Maritime Industry Fundamentals
Maritime Pipefitting
Maritime Structural Fitter

Green/Sustainable Construction

Building Auditor
Fundamentals of Weatherization
Introduction to Weatherization
Sustainable Construction Supervisor
Weatherization Crew Chief
Weatherization Technician
Your Role in the Green Environment

Energy

Alternative Energy
Introduction to the Power Industry
Introduction to Solar Photovoltaics
Introduction to Wind Energy
Power Industry Fundamentals
Power Generation Maintenance
   Electrician
Power Generation I&C Maintenance
   Technician
Power Generation Maintenance
   Mechanic
Power Line Worker
Power Line Worker: Distribution
Power Line Worker: Substation
Power Line Worker: Transmission
Solar Photovoltaic Systems Installer
Wind Turbine Maintenance
   Technician

Pipeline

Control Center Operations, Liquid
Corrosion Control
Electrical and Instrumentation
Field Operations, Liquid
Field Operations, Gas
Maintenance
Mechanical

Safety

Field Safety
Safety Orientation
Safety Technology

Supplemental Titles

Applied Construction Math
Careers in Construction
Tools for Success

Management

Fundamentals of Crew Leadership
Project Management
Project Supervision

Spanish Titles

Acabado de concreto: nivel uno,
   nivel dos
Aislamiento: nivel uno, nivel dos
Albañilería: nivel uno
Andamios
Aparejamiento básico
Aparajamiento intermedio
Aparajamiento avanzado
Carpintería:
   Introducción a la carpintería,
   nivel uno; Formas para
   carpintería, nivel tres
Currículo básico: habilidades
   introductorias del oficio
Electricidad: nivel uno, nivel dos,
   nivel tres, nivel cuatro
Encargado de señales
Especialista en aplicación de
   revestimientos industriales: nivel
   uno, nivel dos
Herrería: nivel uno, nivel dos, nivel
   tres
Herrería) de refuerzo: nivel uno
Instalación de rociadores: nivel uno
Instalación de tuberías: nivel uno,
   nivel dos, nivel tres, nivel cuatro
Instrumentación: nivel uno, nivel
   dos, nivel tres, nivel cuatro
Orientación de seguridad
Mecánico industrial: nivel uno, nivel
   dos, nivel tres, nivel cuatro, nivel
   cinco
Paneles de yeso: nivel uno
Seguridad de campo
Soldadura: nivel uno, nivel dos,
   nivel tres

Portuguese Titles

Currículo essencial: Habilidades
   básicas para o trabalho
Instalação de encanamento
   industrial: nível um, nível dois,
   nível três, nível quatro

Acknowledgments

This curriculum was revised as a result of the farsightedness
and leadership of the following sponsors:

Becon

Cianbro

Jacobs

TIC/The Industrial Company

Zachry Construction Corporation

This curriculum would not exist were it not for the dedication and
unselfish energy of those volunteers who served on the Authoring Team.
A sincere thanks is extended to the following:

Glynn Allbritton

Tina Goode

Paul LaBorde

Ed LePage

Rick Rankin

NCCER PARTNERING ASSOCIATIONS

American Fire Sprinkler Association
Associated Builders and Contractors, Inc.
Associated General Contractors of America
Association for Career and Technical Education
Association for Skilled and Technical Sciences
Carolinas AGC, Inc.
Carolinas Electrical Contractors Association
Center for the Improvement of Construction Management and Processes
Construction Industry Institute
Construction Users Roundtable
Construction Workforce Development Center
Design Build Institute of America
GSSC – Gulf States Shipbuilders Consortium
Manufacturing Institute
Mason Contractors Association of America
Merit Contractors Association of Canada
NACE International
National Association of Minority Contractors
National Association of Women in Construction
National Insulation Association
National Ready Mixed Concrete Association
National Technical Honor Society
National Utility Contractors Association
NAWIC Education Foundation
North American Technician Excellence
Painting & Decorating Contractors of America
Portland Cement Association
SkillsUSA®
Steel Erectors Association of America
U.S. Army Corps of Engineers
University of Florida, M. E. Rinker School of Building Construction
Women Construction Owners & Executives, USA

Pipefitting Level Four

08401-07

Advanced Blueprint Reading

08401-07
Advanced Blueprint Reading

Topics to be presented in this module include:

1.0.0	Introduction	1.2
2.0.0	Piping and Instrumentation Drawings	1.2
3.0.0	Piping Arrangement Drawings	1.8
4.0.0	Reading and Interpreting P&IDs and Piping Arrangement Drawings	1.12
5.0.0	Reading and Interpreting Isometric Drawings	1.14
6.0.0	Following a Single Line	1.16
7.0.0	Drawing ISOs	1.30

Overview

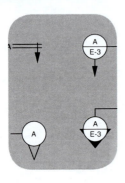

This module teaches the trainee to work through a set of drawings and extract the information necessary from one level to another. This includes using plan views to draw isometrics and using isometrics to put together spools. The drawings start with a site plan for a power plant and follow the process of understanding the set of plans by working through the successive levels. After the site plan, the trainee examines general arrangement plans for a heat recovery steam generator in the system. The trainee then goes from a general arrangement piping plan to the isometrics for one line and the spool drawings associated with that line. The drawings supplied fit together to design a main steam line for a power plant.

Objectives

When you have completed this module, you will be able to do the following:

1. Identify symbols and abbreviations on P&IDs.
2. Identify piping arrangement drawings.
3. Read and interpret GPS coordinates, control points, and elevation.
4. Read and interpret P&IDs, plan views, and section views.
5. Identify isometric drawings.
6. Read isometric drawings taken from plan views.
7. Draw isometric drawings.
8. Read and interpret spool drawings taken from isometric drawings.

Trade Terms

Dimension   Title block
Revision

Required Trainee Materials

1. Pencil and paper
2. Appropriate personal protective equipment
3. Blueprint package (nine drawings) included with this module

Prerequisites

Before you begin this module, it is recommended that you successfully complete *Core Curriculum*; *Pipefitting Level One*; *Pipefitting Level Two*; and *Pipefitting Level Three*.

This course map shows all of the modules in the fourth level of the Pipefitting curriculum. The suggested training order begins at the bottom and proceeds up. Skill levels increase as you advance on the course map. The local Training Program Sponsor may adjust the training order.

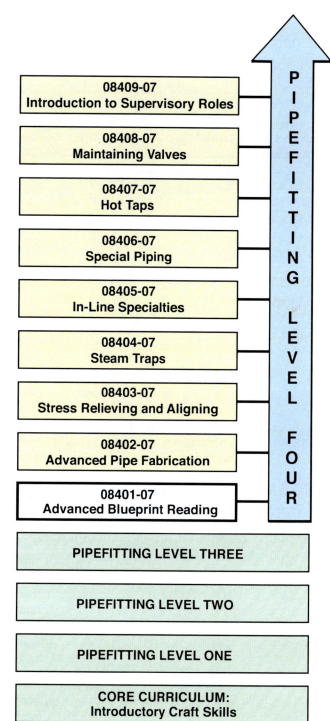

1.0.0 ♦ INTRODUCTION

The purpose of this module is to further your understanding and knowledge of blueprint reading. Building on your previous knowledge of types of drawings, this module explains P&IDs, piping arrangement drawings, and isometric drawings in more depth and explains how to read and interpret these drawings. Control points, coordinates, elevations, and material take-outs are also explained. The blueprint package included with this module contains nine 11" × 17" plans to help demonstrate how information is conveyed and transferred.

1.1.0 Site Plan

Blueprint One is a typical site plan. The site plan for the sample job shows the major parts of a power plant, including the steam plant, tanks, heat recovery systems, and various administrative and storage buildings. The site plan is something like a map of the site. This is where the actual plans start. The site plan is frequently the second page of the set after the title page. The two objects labeled Unit 1 and Unit 2 are heat recovery steam generators (HRSG). Unit 3 is a third generator of a different type, also producing power through a steam turbine. Notice the point at the northwest corner of the site, in the middle of the road, at the intersection of two lines. The location of the point is given as N 1675.97; E 231.13. The site and equipment is laid out with a surveyor's total station based on this point and other points surveyed from this point.

2.0.0 ♦ PIPING AND INSTRUMENTATION DRAWINGS

Piping and instrumentation drawings, or P&IDs, are schematic diagrams of a complete system or systems. P&IDs show the process flow and function of a process. They also show all equipment, pipelines, valves, instruments, and controls necessary for the operation of a particular system. Because their purpose is only to provide a representation of the work to be done, P&IDs are not drawn to scale or dimensioned. Valves and other parts are usually indicated with symbols.

Although a person may remember many of the common symbols and abbreviations used on a P&ID, it is impossible to memorize all of the symbols. There will also be minor differences in the symbols and abbreviations used on different engineers' and contractors' drawings. A legend showing the symbols and abbreviations and the meaning of each often appears in a space on the working diagram or on a separate diagram. *Table 1* shows abbreviations commonly used on P&IDs.

P&IDs provide the following information about a system:

- Process piping
- Piping components
- Process equipment
- Instrumentation

2.1.0 Process Piping

P&IDs show the entire piping system of a process and provide information about each pipeline through the use of lines, line numbers, and match lines.

2.1.1 Lines

The lines on a P&ID show the process flow through a piping system from one place or piece of equipment to another. Each type of line on a P&ID is used for a special purpose. The way that a line is drawn provides the pipefitter with information about that line. *Figure 1* shows types of lines used on P&IDs.

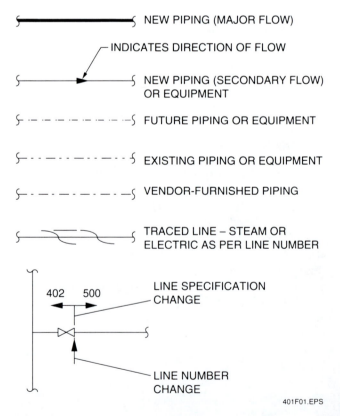

Figure 1 ♦ Lines.

Table 1 Abbreviations (1 of 3)

Adapter	ADPT
Air preheating	A
American Iron and Steel Institute	AISI
American National Standards Institute	ANSI
American Petroleum Institute	API
American Society for Testing and Materials International	ASTM
American Society of Mechanical Engineers	ASME
Ash removal water, sluice or jet	AW
Aspirating air	AA
Auxiliary steam	AS
Benchmark	BM
Beveled	B
Beveled end	BE
Beveled, both ends	BBE
Beveled, large end	BLE
Beveled, one end	BOE
Beveled, small end	BSE
Bill of materials	BOM
Blind flange	BF
Blowoff assembly	BO
Bottom of pipe	BOP
Butt weld	BW
Carbon steel or cold spring	CS
Cast iron	CI
Ceiling	CLG
Chain operated	CO
Chemical feed	CF
Circulating water	CW
Cold reheat steam	CR
Compressed air	CA
Concentric or concrete	CONC
Condensate	C
Condenser air removal	AR
Continue, continuation	CONT
Coupling	CPLG
Detail	DET
Diameter	DIA
Dimension	DIM
Discharge	DISCH
Double extra-strong	XX STRG
Drain	DR
Drain funnel	DF
Drawing	DWG
Drip leg or dummy leg	DL
Ductile iron	DI
Dust collector	DC
Each	EA
Eccentric	ECC
Elbolet	EOL
Elbow	ELB
Electric resistance weld	ERW
Elevation	EL
Equipment	EQUIP
Evaporator vapor	EV
Exhaust steam	E
Expansion	EXP
Expansion joint	EXP JT
Extraction steam	ES
Fabrication (dimension)	FAB
Face of flange	FOF
Faced and drilled	F&D
Factory Mutual	FM
Far side	FS
Feed pump balancing line	FB
Feed pump discharge	FD
Feed pump recirculating	FR
Feed pump suction	FS
Female	F
Female	FM
Female pipe thread	FPT
Field support or forged steel	FS

Table 1 Abbreviations (2 of 3)

Field weld	FW
Figure	FIG
Fillet weld	W
Finish floor	F/F
Finish floor	FIN FL
Finish grade	FIN GR
Fitting	FTG
Fitting makeup	FMU
Fitting to fitting	FTF
Flange	FLG
Flat face	FF
Flat on bottom	FOB
Flat on top	FOT
Floor drain	FD
Foundation	FDN
Fuel gas	FG
Fuel oil	FO
Gauge	GA
Galvanized	GALV
Gasket	GSKT
Grating	GRTG
Hanger	HGR
Hanger rod	HR
Hardware	HDW
Header	HDR
Heat traced, heat tracing	HT
Heater drains	HD
Heating system	HS
Heating, ventilating, and air conditioning	HVAC
Hexagon	HEX
Hexagon head	HEX HD
High point	HPT
High pressure	HP
Horizontal	HORIZ
Hot reheat steam	HR
Hydraulic	HYDR
Increaser	INCR
Input/output	I/O
Inside diameter	ID
Insulation	INS
Invert elevation	IE
Iron pipe size	IPS
Isometric	ISO
Issued for construction	IFC
Lap weld	LW
Large male	LM
Length	LG
Long radius	LR
Long tangent	LT
Long weld neck	LWN
Low pressure	LP
Low-pressure drains	DR
Low-pressure steam	LPS
Lubricating oil	LO
Main steam	MS
Main system blowouts	BL
Makeup water	MU
Male	M
Malleable iron	MI
Manufacturer	MFR
Manufacturer's Standard Society	MSS
National Pipe Thread	NPT
Nipolet	NOL
Nipple	NIP
Nominal	NOM
Not to scale	NTS
Nozzle	NOZ
Outside battery limits	OSBL
Outside diameter	OD
Outside screw and yoke	OS&Y
Overflow	OF
Pipe support	PS
Pipe tap	PT
Piping and instrumentation diagram	P&ID
Plain	P
Plain end	PE
Plain, both ends	PBE
Plain, one end	POE
Point of intersection	PI
Point of tangent	PT
Pounds per square inch	PSI
Process flow diagram	PFD
Purchase order	PO
Radius	RAD
Raised face	RF
Raised face slip-on	RFSO

Table 1 Abbreviations (3 of 3)

Raised face smooth finish	RFSF	Temperature or temporary	TEMP
Raised face weld neck	RFWN	That is	I.E.
Raw water	RW	Thick	THK
Reducer, reducing	RED	Thousand	M
Relief valve	PRV-PSV	Thread, threaded	THRD
Ring-type joint	RTJ	Threaded	T
Rod hanger or right hand	RH	Threaded end	TE
		Threaded, both ends	TBE
Safety valve vents	SV	Threaded, large end	TLE
Sanitary	SAN	Threaded, one end	TOE
Saturated steam	SS	Threaded, small end	TSE
Schedule	SCH	Threadolet	TOL
Screwed	SCRD	Top of pipe or top of platform	TOP
Seamless	SMLS	Top of steel or top of support	TOS
Section	SECT	Treated water	TW
Service and cooling water	SW	Turbine	TURB
Sheet	SH	Typical	TYP
Short radius	SR		
Slip-on	SO	Underwriters Laboratories	UL
Socket weld	SW		
Sockolet	SOL	Vacuum	VAC
Stainless steel	SS	Vacuum cleaning	VC
Standard weight	STD WT	Vents	V
Steel	STL	Vitrified tile	VT
Suction	SUCT		
Superheater drains	SD	Wall thickness or weight	WT
Swage	SWG	Weld neck	WN
Swaged nipple	SN	Weldolet	WOL
		Well water	WW
		Wide flange	WF

2.1.2 Line Numbers

Line numbers are used on P&IDs to indicate the size of pipe, service of line, piping material specification group number, number of the pipe in a particular service, and insulation required. In other P&IDs or other types of drawings of the same system, line numbers are used to identify lines and to transfer pipelines from one drawing to another.

Line numbers are sometimes enclosed in an oval-shaped symbol with an arrow at one end. The arrow indicates the direction of flow through the pipeline and should be used as often as necessary to maintain continuity. The line number can be shown directly on the line or above the line, with an arrow pointing to the line. *Figure 2* shows typical line numbers.

In *Figure 2*, 6" indicates the nominal pipe size; LW is the fluid symbol and represents white liquor in this case; the number 20 indicates the piping material specification; the number 12 indicates the particular line number in the white liquor system; and the letter A indicates the insulation group numbers from the specifications. If no insulation letter is given, insulation is not required. Insulation may be presented in a separate symbol.

Line numbers vary between different engineering companies, and you must always check the legend of the P&ID to determine the system for reading the line numbers. The line number may include the pipe size in the line code.

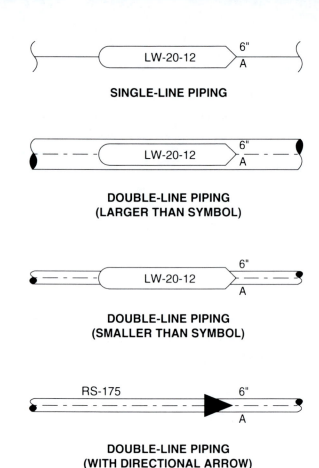

Figure 2 ♦ Line numbers.

2.1.3 Match Lines

Reading a P&ID is a simple matter of tracing a system from one location or piece of equipment to another and interpreting the symbols and abbreviations on the diagram. Sometimes, an entire system is shown on a single sheet, which makes it very easy to trace the system. Larger systems may require the use of two or more sheets to show the entire system, and the system must be traced from one P&ID to another. Match lines help you do this.

A match line (*Figure 3*) is provided at the end of the P&ID to show where the system leaves one numbered sheet and starts again on another numbered sheet. The system is traced to the numbered match line on one sheet and continues at the same number on the match line of the next sheet. The system always continues on the next P&ID in the numbering sequence of P&ID sheets unless otherwise noted on the match line. If the system continues on a P&ID sheet other than the next sheet in sequence, a note to that effect appears in a box on the system line at the match line.

2.2.0 Piping Components

P&IDs show piping system components, such as valves, strainers, expansion joints, clean-outs, flexible connections, dampers, desuperheaters, removable spools, drains, and vents. P&IDs use symbols and abbreviations to describe these components. Always check the legend on the drawings to determine exactly what the symbols stand for. These symbols differ between engineering companies. *Figure 4* shows some commonly used piping component symbols.

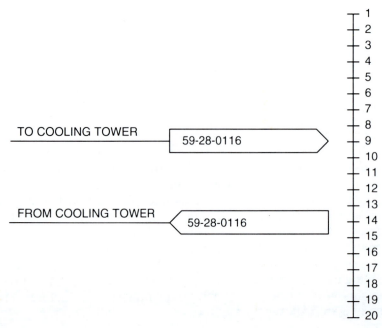

Figure 3 ♦ Match lines.

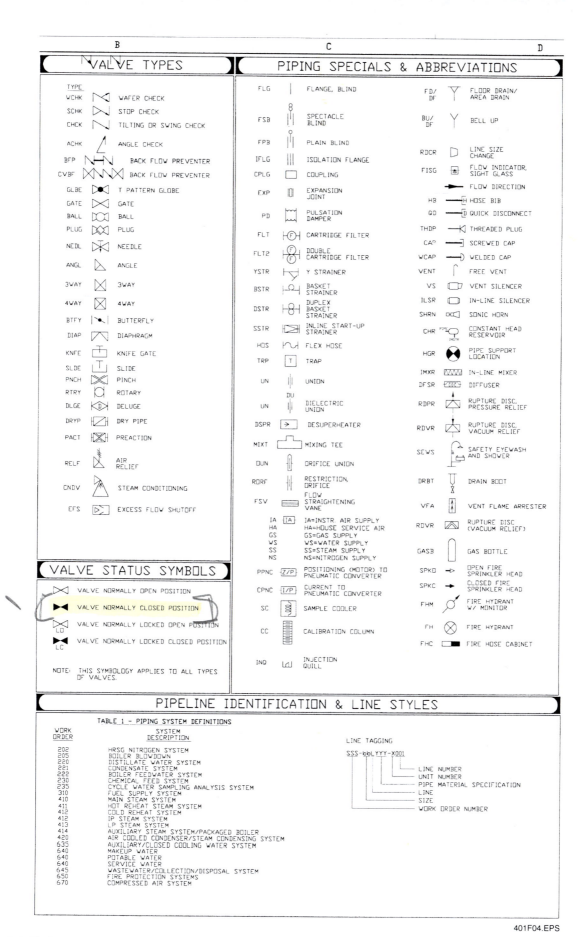

Figure 4 ♦ Piping component symbols.

2.3.0 Process Equipment

P&IDs show process equipment, such as pumps, vessels, motors, turbines, condensers, and other major equipment in a system. P&IDs also give information about the size, capacity, horsepower, and operating parameters of the process equipment. P&IDs often show an area that is enclosed by a dashed line and has the initials VS (vendor's supply), or a vendor's name inside the enclosed area. The VS indicates that all of the equipment within the enclosed area is vendor-supplied; therefore, you are only responsible for the piping up to that area. *Figure 5* shows commonly used equipment symbols. These symbols may be different on different projects, depending on the engineers.

2.4.0 Instrumentation

P&IDs show system instrumentation, such as control valve operators and analytical, flow, pressure, level, and temperature instruments. P&IDs use symbols and abbreviations to describe the instruments and the functions in a system. *Figure 6* shows typical instrumentation symbols and the meaning of each.

3.0.0 ◆ PIPING ARRANGEMENT DRAWINGS

Piping arrangement drawings show how the equipment is arranged and the locations of all equipment used with the system. Piping arrangement drawings can be either plan or section views. Equipment on these drawings is usually drawn to correspond to the basic shape of the item represented. These representations are common to most companies.

A plan view is a view of an object or area from above, looking down on the system. The plan view shows the exact locations of the equipment in relation to established reference points, objects, or surfaces. These reference points may be building columns, walls, benchmarks, or other equipment.

Plan views are usually laid out with plant north at the top of the drawing. If the need for space makes it necessary to rotate the drawing, it is rotated so that north is to the left side of the sheet. *Figure 7* shows a plan view of a heat recovery steam generator (HSRG).

Sections, or sectional views, are projections of a plane within an object or system. Section views are detailed drawings showing a side or plan view of a particular part of a plan drawing. A section view shows what the internal characteristics look like on the cutting plane. A cutting plane may not be continuous, but a horizontal cutting plane of

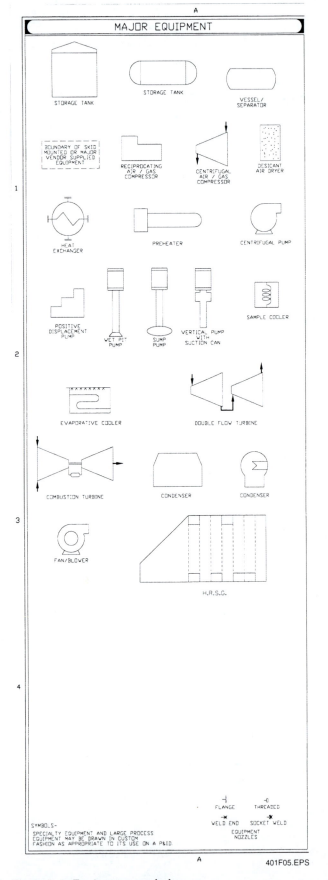

Figure 5 ◆ Equipment symbols.

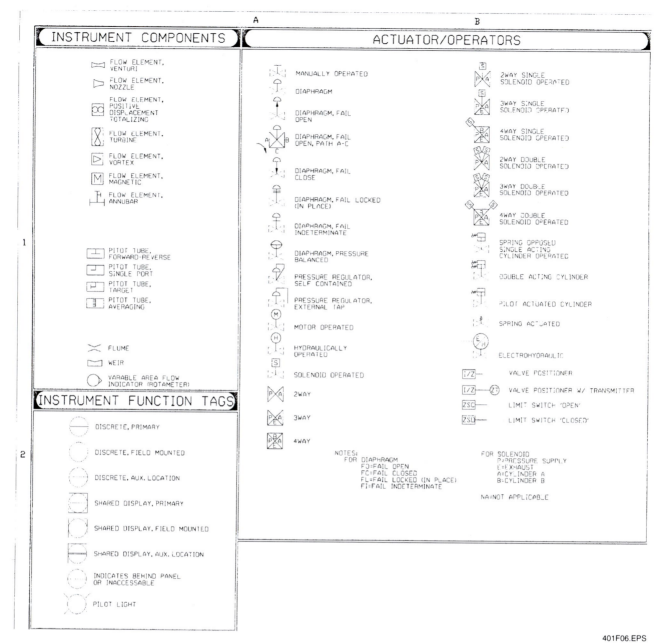

Figure 6 ♦ Instrumentation symbols.

a building may be staggered to include specific information.

Because of the way section views are drawn, they provide visual information not included on elevations or plan views. Sections are sometimes a cutaway view of an object to show internal arrangements that may not be obvious on other views. There are many symbols used on drawings to designate section cutting planes. *Figure 8* shows common symbols used for section views.

Blueprint Six shows section views taken from a power plant. The three section views of headers C, D, and E, as well as detail G, are taken from the section view of an elevation of an HRSG.

Piping arrangement drawings use control points, coordinates, and elevations to indicate the exact location that a pipeline, floor penetration, or piece of equipment is to be placed in a building. Elevations refer to the vertical placement of an object; control points and coordinates refer to the horizontal locations. When identifying the location of an object in a building or area, first locate the horizontal location, and then establish the elevation.

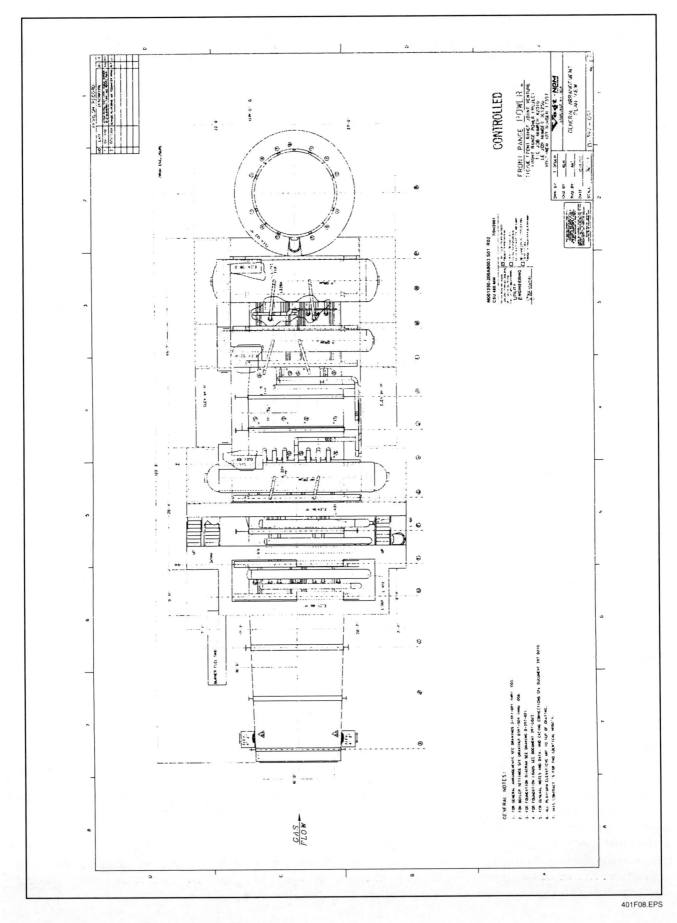

Figure 7 ♦ Plan view of heat recovery steam generator.

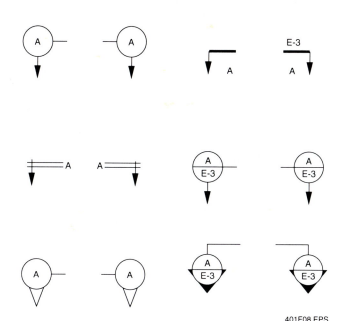

Figure 8 ◆ Symbols for section views.

3.1.0 Control Points

Column line control points are used to locate the work horizontally. The nearest column line on a drawing is shown, and the piping is then dimensioned from this column. This column is designated using the grid line system. The grid line system designates every column inside a building or structure. On a grid line drawing, an imaginary grid is made using the column lines. The areas within the column lines are known as bays. *Figure 9* shows a grid line drawing.

The numbering and lettering of the system normally begins in the upper left- or right-hand corner of the system, which is the north side of the building or structure. The numbers are normally assigned to the vertical rows, and the letters are assigned to the horizontal rows. In some cases, the letters I, O, and Q and the numbers 1 and 0 have been eliminated from the system to avoid confusion. This system allows each beam in the building to be given a reference number or control point, such as B-3. Normally, the engineer labels each beam in a building for ease in identification. The use of column lines permits anyone who is familiar with the system to locate any point exactly. When column lines are used with elevations, the locations can be specified three-dimensionally.

3.2.0 Coordinates

Coordinates are a system of numbers used to define the geographic position of a point, line, piece of equipment, or other object on the plant site. A coordinate is the distance a given point is from a base point, commonly called a benchmark (BM). Coordinates are written as a letter and number combination. The letter designates the direction of the coordinate, normally north or east, in relation to the benchmark, and the number designates the distance in feet and inches. Coordinates are used to locate major equipment on drawings.

A site plan indicates the position of the originating point or coordinate benchmark. From this point, a grid system is developed and is used by all design groups to coordinate their drawings. The north-south axis and the east-west axis intersect through this originating point, and every point on the drawing can be referenced by the originating

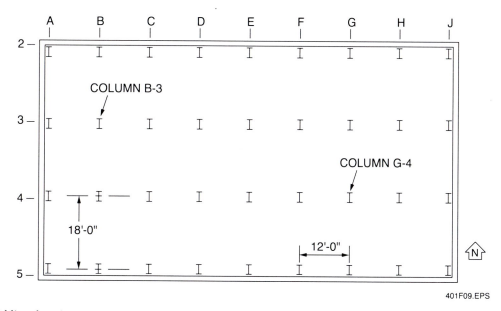

Figure 9 ◆ Grid line drawing.

point. For example, a point that is 30 feet east of the north-south axis through the originating point at N. 1675.97 and E. 231.13 has an east coordinate of E. 261.13. If this same point were 1 foot north of the east-west axis through the originating point, it would have a north coordinate of N. 1676.97. Put together, the two coordinates pinpoint the location of the point as E. 261.13 and N. 1676.97. *Figure 10* shows coordinate origin. Frequently, the elevation is given with the coordinates.

3.3.0 Elevations

Elevations locate piping vertically in relation to other runs of pipe, building floors, or platforms. Elevations may be referenced to the center line (CL) of the pipe, to the bottom of a pipe, to the face of a flange, or to any other convenient point. The drawing must indicate which point the reference is made to and must always be consistent. Engineers normally mark elevations on the structural columns using a reference line. These elevations are written in feet and inches. To determine the exact elevation at which a pipeline must be set, locate the elevation reference point marked by the engineer.

Piping elevations are specified in different ways. Sometimes, the elevation is given as an above-the-finished-floor elevation, in which case you would measure from the finished floor to the desired height to establish the equipment elevation. In other cases, the elevation must be determined from an established elevation benchmark.

An elevation benchmark may be inside or outside the building. Usually, the benchmark is located on building columns in various places throughout the building. To determine a rough elevation from a benchmark on a column, you can measure from the benchmark to the floor, using the floor as a measuring point. For a more accurate measurement, more precise methods must be used.

Elevations can be transferred from the benchmark to the bottom of the pipe or face of the flange, using a string level, an optical level, a theodolite or total station. The most precise tool is the laser level.

4.0.0 ◆ READING AND INTERPRETING P&IDs AND PIPING ARRANGEMENT DRAWINGS

P&IDs provide an overall control scheme of the piping on a project. They show the types of control mechanisms and how those control mechanisms are used. They show the number of block valves and instruments, identify line numbers and equipment, indicate flow direction, and give other detailed information required on the project. Plan drawings show a view of an object or area from above, looking down on the system,

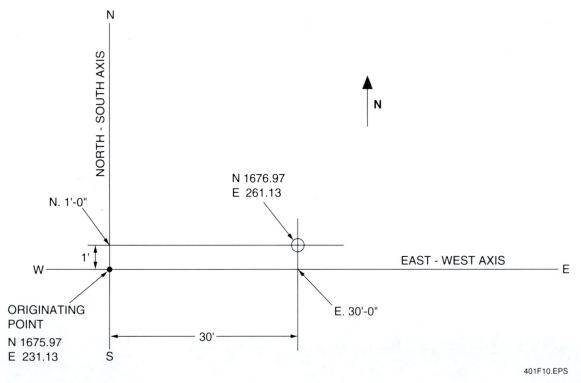

Figure 10 ◆ Coordinate origin.

and show the exact locations of the equipment and piping in relation to established reference points, objects, or surfaces. A section view is the result of an imaginary line cut through the system. From this drawing, the height, width, and location of the equipment can be seen.

In this module, drawings are used to teach you how to read and interpret P&IDs, ISOs, plan drawings, and spool drawings. This section teaches you how to trace a pipeline through these various types of drawings.

The P&ID shows, in schematic fashion, the interconnected pipelines of a system. *Blueprint Three* shows the main steam system as it comes off the two HRSG units. This system feeds steam into the turbine, which, in turn, drives the generator. The steam supply lines from HRSG Units 1 and 2 feed to lines 12LKEA-1031 and 12LKEA-2031. The two 12-inch lines come together, through 16 × 12 concentric reducers, into a single 16-inch line, which passes through a 16 × 14 concentric reducer into the steam turbine. Eleven smaller lines or instruments come off the 16-inch line. Two of the lines go to the steam turbine area flash tanks. The 16-inch line run has instrumentation shown and identified, including temperature and pressure indicators and primary elements.

Starting at the intersection of the two 12-inch lines, there are a series of branches to various aspects of the main steam system. The first branch is an 8-inch line, labeled 8LKEA-3064. This line runs to a drain through a pair of normally closed globe valves and a welded cap leading to the drain. The other line from the 8-inch pipe is stepped down from 8 inches to 2 inches, where it connects to an automatic balancing valve pneumatically controlled by temperature sensors in a thermowell on the 8-inch line. The end of this 2-inch line goes to the steam turbine flash tank. On the ISO, and on the spool sheet for the section of 16-inch pipe that the 8-inch line comes from, the 8-inch line is only shown for 2 feet.

Returning to the 16-inch line, the next connection is a vent line, normally closed off by the two ball valves on it. After this branch, the hexagonal symbol with C.S. and 5.5 on it tells the fitter that the line is insulated with 5½ inches of calcium silicate, based on the Piping and Miscellaneous Tags section of the mechanical symbology sheet. The line continues with four sensor and transmitter lines, three pressure transmitter/indicator combinations and a temperature sensor, transmitter and indicator. The pressure transmitters are isolatable by means of ball valves.

Another 8-inch line comes off the main steam line, and another balancing valve assembly transfers steam to the steam turbine flash tanks. The thermal control assemblies on these allow the operating temperature to control the delivery of steam, as a safety process.

Two test connections are connected to the main line, followed by two thermowells, with or without temperature probes. Finally, the P&ID shows the 16-inch pipe stepping down to 14 inches, raising the speed of the steam as it enters the steam turbine. The turbine is shown as being installed and set up by the turbine manufacturer.

4.1.0 Mechanical Symbology Page

The mechanical symbology page (*Blueprint Four*) shows the standard symbols used on this set of drawings by this engineering company. Separate sections show the abbreviations, the symbols for valves, valve status, and specials, and the codes for class and specifications for different components. The explanations for different line numbers and pipe identification numbers is also shown on this page. You will need to be able to refer to this page when you are working with this set. For example, there is a set of codes and descriptions that are added to a line tag, (seen in the lower left of the figure), which tells you that a pipe line with that abbreviation on the tag is to be that specified pipe. The various valve symbols will appear again and again. Notice the section labeled Valve Status Symbols, which identifies the normal condition of valves as set. If a valve is shown as normally open, you would obviously not attempt to disassemble the system downstream of that valve. The section labeled Equipment/Valves/Specials Function provides codes identifying the functions of various components. The section for Piping and Miscellaneous Tags identifies drains, valve tags, field tags for various pipe components, and insulation specifications.

4.2.0 Instrumentation Symbology Page

The instrumentation symbology page (*Blueprint Five*) has the symbols for instrumentation and for operators and actuators. The categories are Components, Functions Tags, and Line Styles for Instruments, Instrument Function Abbreviations, and Actuators and Operators for Various Equipment. An example of a component is a flow element, which is one of several varieties of flow meters or sensors that controls either a readout or a process, or both. An instrument that is identified as a discrete, primary component will have a set of initials inside the circle. These initials are a code referenced in the section titled Instrument Function. If the code in the circle was TI, the instrument would be a temperature indicator. If the tag read PT, that would be a blind pressure transmitter.

4.3.0 General Arrangement Pages

The general arrangement pages show elevations (*Blueprint Six*) and plan views (*Blueprint Seven*) for the heat recovery steam generators, Units 1 and 2 from the site. Ladders and stairwells are shown, as are the high-pressure, intermediate-pressure, and low-pressure drums, and external piping and ducts. The external structure is apparent also. Here the piping is not labeled in any detail, only drawn as it is located on the HRSG. Several other general arrangement pages were supplied with the drawing set, showing the piping arrangements at different heights within the system. An entire set of prints would normally be drawn showing the layout of underground and aboveground piping on the site.

5.0.0 ♦ READING AND INTERPRETING ISOMETRIC DRAWINGS

This module takes a closer look at ISOs and the information they contain. A pipefitter in the field must have a thorough knowledge of ISOs.

On an ISO, all horizontal lines are projected at 30 degrees. Vertical lines remain vertical. Dimensioning is the same as on other types of piping drawings. The overall effect of an ISO is the perception of depth. An ISO shows the height, width, and depth of a pipeline, making it easier to visualize sections of pipe.

ISOs are widely used in the fabrication and erection of pipe. The biggest advantage that ISOs have over other types of drawings is clarity. In an ISO, the draftsperson can extend a pipeline and show all its components without regard to scale. Therefore, an ISO can show a complete line from one piece of equipment to another or show a complete line as it exists on an orthographic drawing. All the information needed to erect and fabricate the line can be clearly shown on an ISO. In addition, the ISO usually contains a bill of material in which all components are listed. Some companies fabricate directly from ISOs; other companies fabricate from spool sheets that are taken from the ISOs. All companies, however, make ISOs from orthographic drawings, and pipefitters use ISOs frequently in the field.

Information that can normally be found on an ISO includes the following:

- North arrow
- Dimensions and angles
- Reference number of the orthographic drawing from which the ISO was taken
- Reference numbers for each spool in the system
- Line numbers
- Direction of flow
- Insulation specifications
- Steam or electrical tracings
- Equipment numbers and location by center line of equipment
- Size and type of valves
- Size, pressure rating, and instrument number of control valves
- Number, location, and orientation for each instrument connection
- Details of flanged nozzles in equipment in which piping is to be connected
- Locations of field welds
- Match line information
- Unions required for installation and maintenance
- Construction materials specifications
- Locations of vents, drains, and traps
- Locations of pipe supports identified by their number
- Special requirements of the system

5.1.0 Isometrics

The general travel of piping is on the isometric drawings of the line. *Figure 11* is an undimensioned isometric of the entire main steam line, giving a rough idea of where it all goes. The sections of straight pipe may or may not be shown on spool drawings. Spool and shop drawings show the run of the pipe, with the connections for tees and laterals shown only at the junction with the pipe. ISOs show the branch connections, at least in part. The valves and instruments are identified on ISOs.

ISOs are not drawn to scale. The descriptions of components and pipe are supplied in the Fabrication Materials list on the upper right side of a typical ISO sheet, with identifying numbers at each component on the sheet. These allow the job to be broken down into sections and distributed to fitter teams. The ISO can then be used as a guide for assembling in order the spools that make up the line.

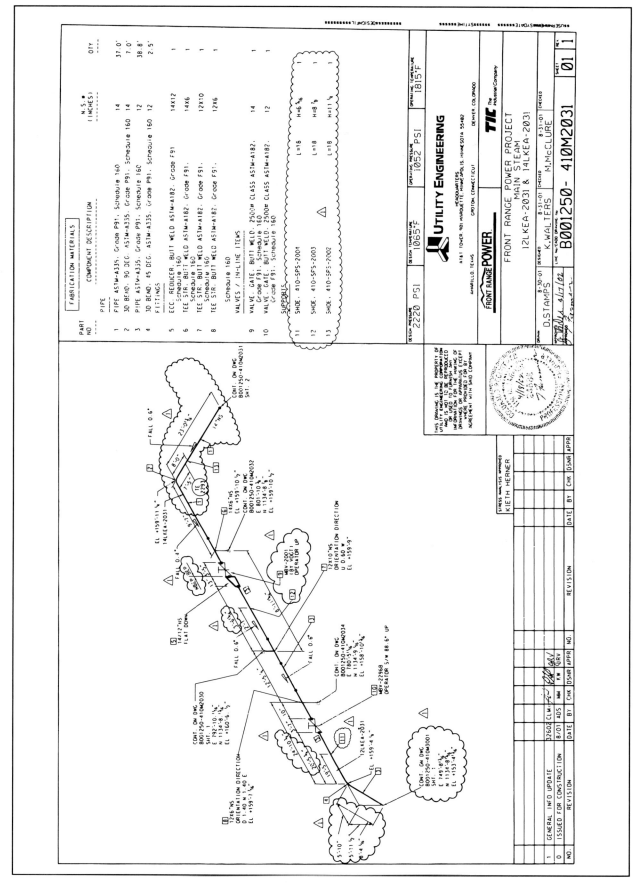

Figure 11 ◆ ISO of part of the main steam system.

5.2.0 Spool Drawings

Spool drawings are specifically used to fabricate sections for the fitters, so that the pieces are manageable, but are put together in right relationship. The spool fabrication may be done in a separate shop, or contracted to a shop. It may also be done by the same team that puts the run together, on or off site. It is much easier to do the majority of the assembly on the floor, and then raise the spool into position and attach it.

The spool sheets show how the 16-inch line develops. As shown on the P&ID (*Blueprint Three*), the two 12-inch lines, 12LKEA1031 and 12LKEA2031 come together and pass through 12 × 16 concentric reducers to form the line 16LKEA3001. Follow the spool drawings to see the spools that must be fabricated to carry the steam to the turbine. At the same time, follow the line on the ISOs, so that you can see the branch lines and what they do, and to see the orientations of the line.

A spool sheet usually has a list of parts, with reference numbers. Piping segments are generally identified in a separate cut list on the sheet, with identifiers. There is sometimes just a bill of materials, without separating individual pieces of pipe.

5.3.0 Vessel or Unit Drawings

You may have vessel or subassembly prints (*Blueprint Eight*) to work from or connect to. This single sheet is ready to send out as a shop drawing, although detailers in the shop may break the sheet down further into specific fabrications. The vessel will be shown on a P&ID and in general arrangement drawings. This sheet shows the necessary details to attach to whatever other components are connected to it.

The vessel itself is a blowdown tank, shown in plan and elevation view. An assembly drawing is provided for the 10-inch drain, which is to be welded to the tank for hydro testing. The 10-inch valve is to be shipped loose to be field routed and welded, as is the pipe and valve for the 2-inch drain. The outlets are to be beveled, ready for welding on delivery.

This particular drawing is for a Canadian fabrication shop. Outlets for the tank are labeled as nozzles, numbered N1 through N16. Included in the list are threaded seats for thermowells, described on the list as temperature wells. This sort of language variation is fairly common in different locations.

6.0.0 ◆ FOLLOWING A SINGLE LINE

Start the line by looking at the P&ID (*Blueprint Three*) and finding the two 12-inch lines from HRSG Unit One and Unit Two. The lines are labeled as 12LKEA1031 and 12LKEA2031. The two lines converge and merge to produce Line 16LKEA3001, which provides steam to the steam turbine, shown as the responsibility of the vendor. The line is shown on the P&ID as if it were a straight line, with branch lines and assemblies attached, and ending after reducing to 14 inches and connecting to the turbine.

Now examine *Blueprint Nine*, which is a piping general arrangement drawing for a section of the piping running from the HRSG Unit One and Unit Two main steam lines. This shows the junction of 12LKEA-2031 and 16LKEA-3001, and shows 12LKEA-1031 running parallel, before the junction. There is a break shown, where 16LKEA-3001 runs out of the level shown, and then it returns to the level to the left of the page. Several more pages of plan views at different heights follow this sheet. Notice on *Figure 12* that 16LKEA comes from the side and centers over the Steam Turbine. The connection shown on *Figure 13* is on the other end, where it becomes a 14-inch line connected to the HRSG.

6.1.0 Getting an ISO from a Plan View

When you are pulling an ISO from a plan view or general arrangement drawing, you need to work in a logical manner. Start from a clear break in the piping and proceed to another clear break, a change. Look up the line in the pipe list (*Figure 14*). The line is neither insulated nor heat traced. In *Figure 15*, the logical place to begin or end an ISO is the flange at either end. The location is given, with reference to the benchmark identified in *Blueprint One*. The center line of the flange face on *Figure 15* is given as East 1733'-7½" and North 1583'-5/16", the center line elevation being given as 92'-6". Since the other end of the run on *Figure 16* is set at the same elevation, and the notes tell us that these are continued as underground piping on another drawing. It is fairly probable that 92'-6" is close to the ground, with just enough room for the flange to be clear of the ground.

It appears that the flange is at a 45-degree angle to the ground because at both ends, there is a 45-degree ell. The first ell is centered at 2'-11 15/16" from the flange. In the spool immediately after the ell, there is a 2-inch line to a pressure control valve followed by a 2 to 1 concentric reducer. It is most likely the ISO would simply place a 2-inch weldolet on the 12-inch line, rather than add the

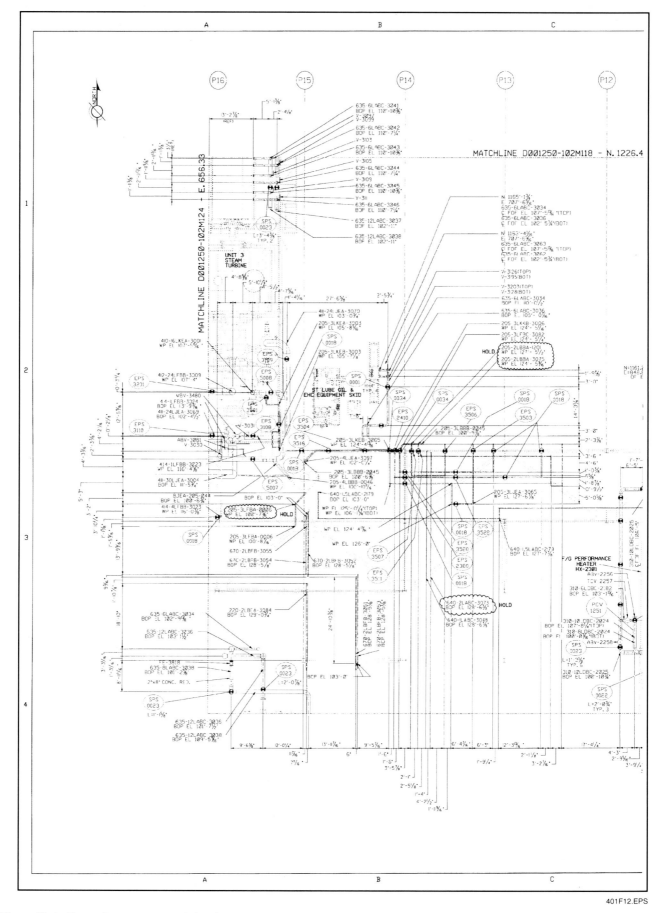

Figure 12 ♦ General arrangement piping 2.

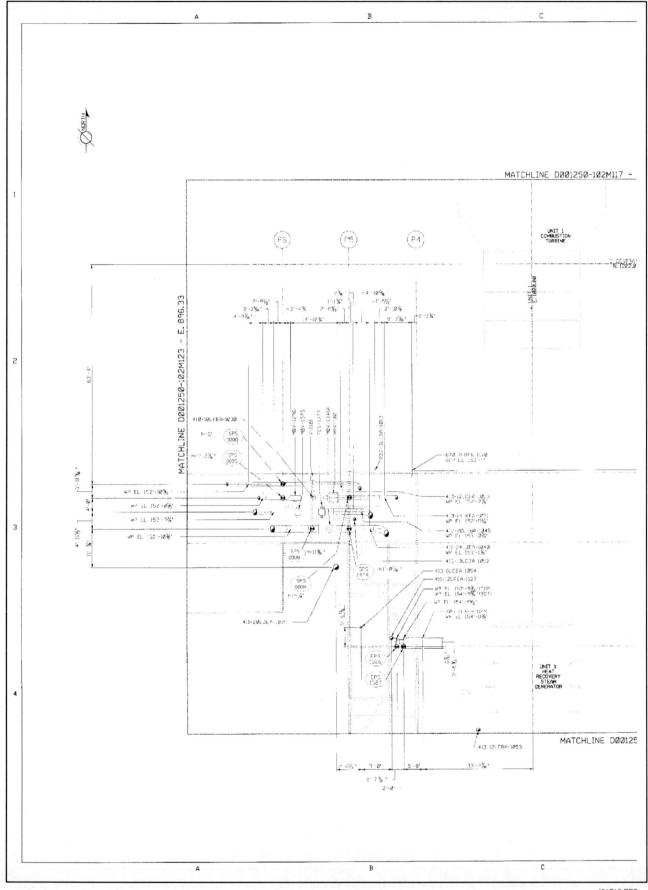

Figure 13 ◆ General arrangement piping 3.

branch. However, it might show the line to the 2 to 1 reducer.

The next item on the line is an ABV, an automated ball valve. This gives control of the flow at this point and is controlled by a pneumatic actuator, which can be controlled remotely. According to the valve list on *Figure 14*, this is a 12 × 10 fail closed valve. This line is a fuel line, as shown by the pipe specifications index description for the number 310 *(Figure 17)*. Look at the column labeled UE Pipe Specification 3-Digit Code and you will find that the FBA in the line label means that this is the main fuel gas line, rated as 300-pound class. In the specification in the valve list, the ABV is referenced to API 6D, which refers to an American Petroleum Institute standard that allows for zero leakage when the valve is closed.

The next item identified on the line is a support, labeled SPS UB24. The specific support will be supplied as a detail, usually from a cutsheet from the manufacturer. These will frequently be provided as several details on a single sheet of the drawings, so that the hangers and support details are all in one place.

Now a spool of 12-inch pipe continues the line, with flow element 0250 in the line. According to the legend for the P&ID, this is a primary flowrate element, a flowrate sensor. The end of the spool is flanged, and is followed by a short spool that has a 12 × 8 concentric reducer welded on the end, ending in a flange. The flange bolts to a PCV, labeled PCV 0251, an 8-inch V-ball pressure control valve. The other side also has a 12 × 8 concentric reducer bolted to the flange of the valve. Again, just before the valve the spool has a support, and after the valve there are three supports and two PRVs, PRV 0089A and 89B. Finally, there is another 45-degree vertical ell and a flanged connection to end the ISO.

To draw the ISO for this, take a piece of ISO paper and start with a 10-inch flange connection, with the elevation of the center line of the flange given as 92'-6", and the line being labeled as a continuation from an underground line. On the ISO, you will identify each component with a small number tag and enter it as an item in the component description list. From there, the line runs up at a 45-degree angle to the bend. The run here is 2'-1½", as is the set. The travel to the center of the bend is 2'-11¹⁵⁄₁₆". Now draw a 45-degree angle and the spool that follows from there to the ABV. In the middle, indicate the 2-inch weldolet, and show a short line to the PCV and the 2 × 1 reducer. Show the 12-inch ball valve symbol and identify it. Now show the spool that follows and the flowmeter. This is a short flanged spool with a measuring device attached. It is a direct mass flow meter, measuring the mass passing through over time. It is to be installed by the instrumentation and control (I&C) division.

The next spool has a 12 × 8 concentric reducer, with the 8-inch end flanged. A support is shown at the point where the 12-inch end of the concentric reducer rests. The next element is PCV 0251. Another 12 × 8 concentric reducer brings the size up to 12 inches again, then a short spool with a pressure relief valve, three more supports, another PRV, then the spool is welded to another 45-degree ell. Last is a spool to the flanged end, with a travel of 4'-2¹³⁄₁₆" from the center of the ell to the flange center line.

6.2.0 From ISO to Spool Sheet

Now take a look at *Figure 18*, which is sheet B001250-410M3001. At the top of the ISO, the line begins with the two converging lines from the ISOs for 12LKEA1031 and 12LKEA2031. There are callouts on the lines telling you that they are continued from the drawings for those lines.

Valve and Pipe Lists

Drawing Number	Pipe Tag	Line	Insulation	Size	Heat Trace	
D001250-310M001 S01 R1	12LFBA--0030	30	N/A	12	N/A	

Drawing Number	Line	Tag	Valve Type	Size	Line Type	Remarks
D001250-310M001 S01 R1	310	ABV-0262	BALL	12×10	FBA	API-6D, N.O., Air Fail Closed
D001250-310M001 S01 R1	310	PCV-0251	V-BALL	8	FBA	By I&C (FC)
D001250-310M001 S02 R1	310	PRV-0089A	ANGLE			HOLD, UE bid
D001250-310M001 S02 R1	310	PRV-0089B	ANGLE			HOLD, UE bid

Figure 14 ◆ Valve and pipe lists.

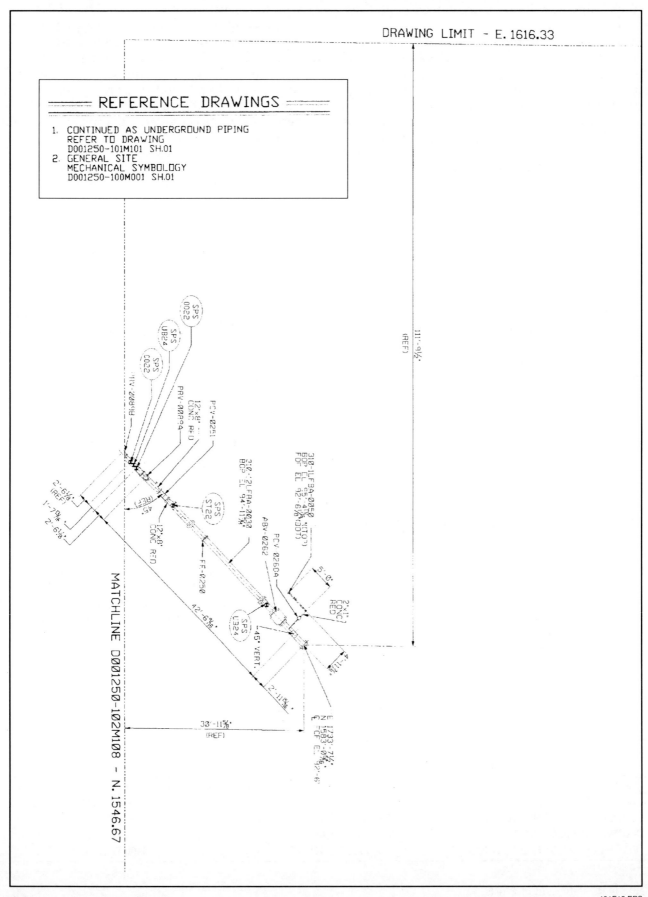

Figure 15 ♦ ISO line LFBA0030.

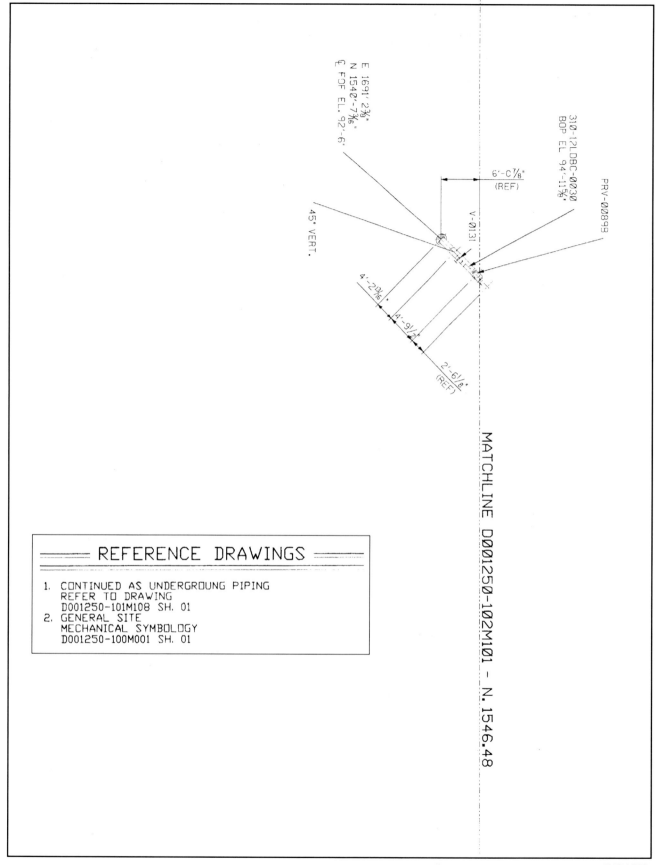

Figure 16 ◆ Continuation of ISO line LFBA0030.

COLORADO SPRINGS – FRONT RANGE POWER PROJECT

Pipe Specification Index

Work Order #	System	ANSI Pipe Class	Design Temperature deg F	Design Pressure psig	Flow lb/hr	Operating Temperature deg F	Operating Pressure psig	Flow lb/hr	UE Pipe Specification 3-Digit Code
205	Boiler Blowdown		660	2,000	45,000/225,000	625	1,900		BFA
220	Distillate	150	100	150		80	110		BBA
221	Condensate Suction	150*	160	25*	1,640,000	130	-5	1,247,823	BBA
221	Condensate, Pump Discharge	300*	160	400*	1,640,000	130	300	1,247,823	DBA
222	Boiler Feedwater Suction	300	335*	200*		310	108		DBA
222	Boiler Feedwater IP Discharge	600*	335*	1,000*		310	700		FBA
222	Boiler Feedwater HP Discharge	1500*	335*	3,000*		315	2,600		JBA
230	Chemical Feed and Sample Panels		1,100	3,000		various	various		TFB/TFD
310	Fuel Gas, Duct Burner	150*	110*	285*		85*	30*		BBC
310	Fuel Gas, Main	300*	110*	550*		85*	460*		DBC/DFA/FBA
410	Main Steam System, Bypass to Cold Reheat	600*	1,065	550*		660	380		JBA
410	Main Steam System, BFP - HP Discharge	1500*	335*	3,000*		315	2,600		KEA
410	HP Steam System	2500*	1,065	2,220		1,052	1,815		DBA
411	Hot Reheat System, Condensate Supply	300*	160	400*		130	300		JEA/FEA
411	Hot Reheat System	1500*	1,065	475		1,052	350		FBB
412	Cold Reheat System	600*	750*	550*		660	395		DBA
413	Low Pressure Steam System, Steam	300	650	150		565	55		DBA
413	Low Pressure Steam System, Condensate Supply	300*	160	400*		130	300		FBB
414	Auxiliary Steam System, Cold Reheat Supply	600*	750*	550*		660	395		
420	Steam Condensing	150*	250	15		121	-10.25		ABC
635	Auxiliary/Closed Cooling Water	150	110*	80*	125,500	85*	31*	71,000	ABC
640	Service Water	150*	100	150		80	115		ABC
640	Softened Water from Nixon Plant	150*	100	150	62,500	80	115	55,000	
645	Waste Water	150*	212	100		60	60		
650	Fire Protection (AG/UG)	175/350	100	175	1,000,000	80	150	500,000	CBA/EDA
670	Compressed Air	150*	150	150		100	125		BFB/BBA

Figure 17 ♦ Piping specification index.

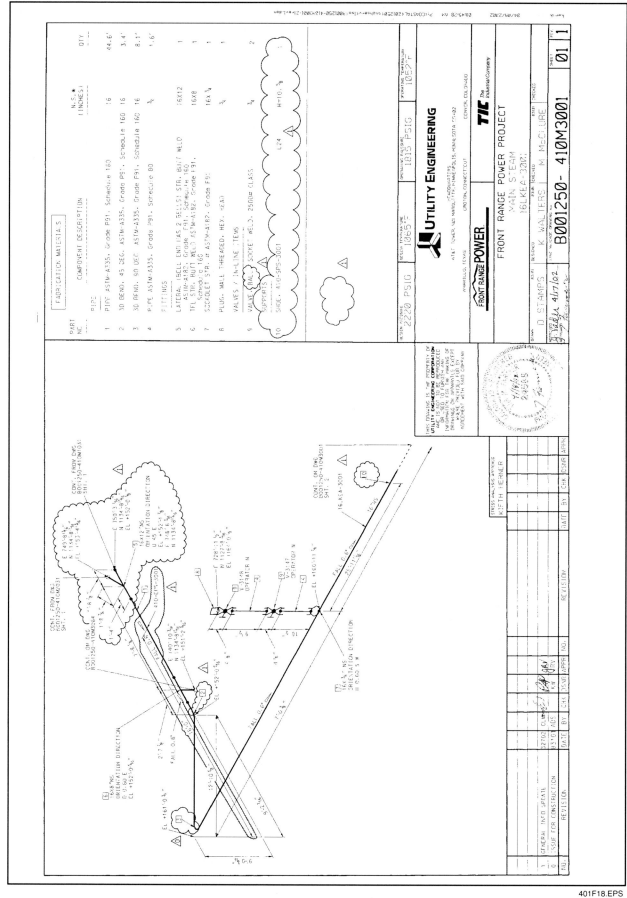

Figure 18 ♦ ISO line 16LKEA-3001 sheet 1.

The cloud shape or curly line around the end area is a way of showing that this area has been revised from an earlier set of drawings. The triangle next to the cloud shape has a 1 in the middle. In the **revision** block the number tells you that on 3/27/02, the general information update was added.

If you examine the changes in the distances given as **dimensions**, you will discover the direction of the pipe run. If the east dimension decreases, the pipe is running west. If the north dimension decreases, the pipe is running south. The locations of the endpoints are given as dimensions from a benchmark, and elevation. These dimensions will be set with a total station using the benchmark, or using marks set from the original benchmark. The spool drawing that matches that beginning is *Figure 19*, which shows the bend and the 12-inch piping coming into the 16 × 12 eccentric reducing lateral where the two lines merge.

In order to keep responsibility clear, it is necessary that the pipefitter has an assignment whose beginning and end points are clear. In this case, the table of materials tells you that you are putting the 12-inch pipe together from the end of the spool to the bend and from there to the lateral. The dimensions are given for the fitup to the 12-inch line, 12LKEA2031, which is from HRSG Unit 2. The system is high-pressure steam, and is made out of Schedule 160 P91 chrome. This information is also shown on the ISO for this section, and the design pressure and operating pressure are shown as well as the design and operating temperature. You will see these, on the ISO, above the **title block**. This is a superheated steam system, as you can tell when you see that the operating pressure is 1,815 psig, and the operating temperature is 1,052°F. The insulation is described by the symbol on the P&ID as 5.5 inches of calcium silicate, to keep that temperature constant.

Welding instructions are given in the right-hand lower corner of the spool sheet. Notice that on these drawings, the welder who butt welds the pipe signs or initials each weld he or she performs. The welds are marked on the spool sheets with a number inside either a hexagon or a diamond that references those signatures. The hexagonal weld symbol is shown as 100 percent X-ray tested, while the diamond is shown as 100 percent magnetic tested.

On the spool sheet, the instructions are given for installing a 1-inch plug on the run at a point 6 inches from the end of the spool. These plugs are installed after the weld has passed the x-ray examination, and the spool is shipped with the plug in place. After the spool is delivered, the pipefitter will install any branch lines shown on the P&ID at that point on the spool.

The next spool *(Figure 20)* has two bends in it, a 45-degree bend and a 90-degree bend. The 45-degree bend is horizontal and the 90-degree bend is vertical. On the tangent of the 45-degree bend is a 16 × 8 reducing tee, starting the 2-foot long 8-inch branch. Observe that the valves on the ¾ line on the ISO, labeled V-3147 and V-3148, are drawn as T-pattern globe valves, but they are drawn on the P&ID as ball valves, and are identified on the ISO component list as ball valves. It is very important that you clarify such questions before you put the line together, as the rework will always be more expensive than doing the job right to start with. It would be easy to say that the engineer changed the words and didn't change the drawing, but make sure. On the spool, the fabricator will install a thermowell 9 inches from the end of the branch. In addition, there is to be a weldolet centered 3 inches from the end, and a cap with a ¾-inch sockolet in it on the end. A detail sheet *(Figure 21)* specifies the welding and fit for the fittings. Notice on the ISO that the pipe is to be installed with a 0.6 fall. This slope is incorporated in the elevations at the points where they are given, but can also be calculated for each piece as it is installed.

The next spool *(Figure 22)* is also on this ISO. Here, the fabricator in the shop installs a 16 × ¾ sockolet 6 inches from the beginning of the spool for the vent assembly shown on the ISO. The assembly to be put together by the fitter will consist of two ¾-inch ball valves as isolation valves, with a threaded plug in the end. Notice that the revision has added item 10, a shoe support.

The next ISO *(Figure 23)* has another shoe support added as a revision, as well as a restrained joint. Because of the large difference in temperature from cold to hot in this process, these are sliding supports and variable can hangers and supports, which allow the expansion and contraction of the pipe to take place without binding. There are details, proprietary to the manufacturer of the supports and hangers, which are supplied with the drawings. These drawings, also called cut sheets, are supplied with the bid documentation to demonstrate compliance with the specifications for the job.

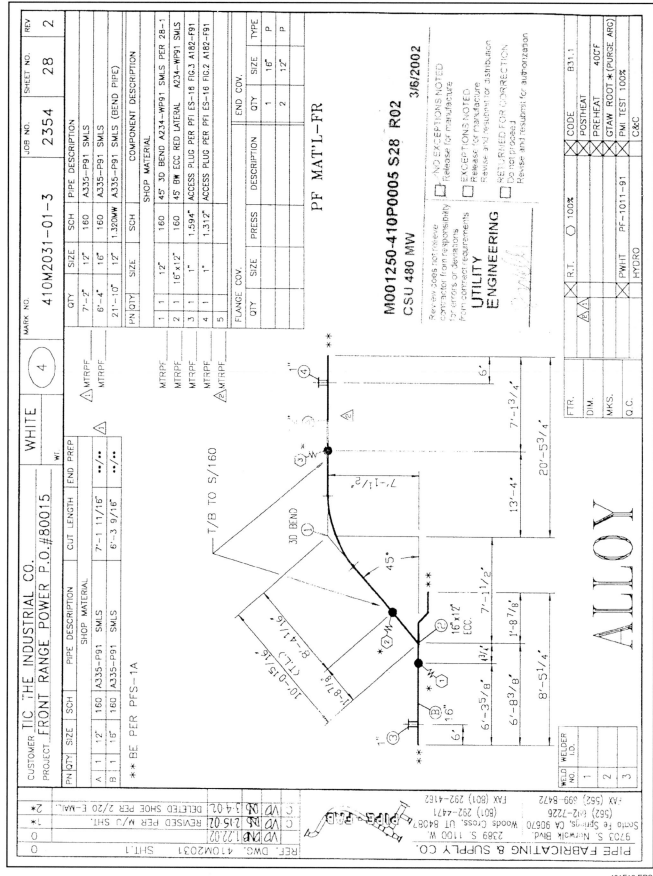

Figure 19 ♦ Spool page 16LKEA-3001 sheet 1.

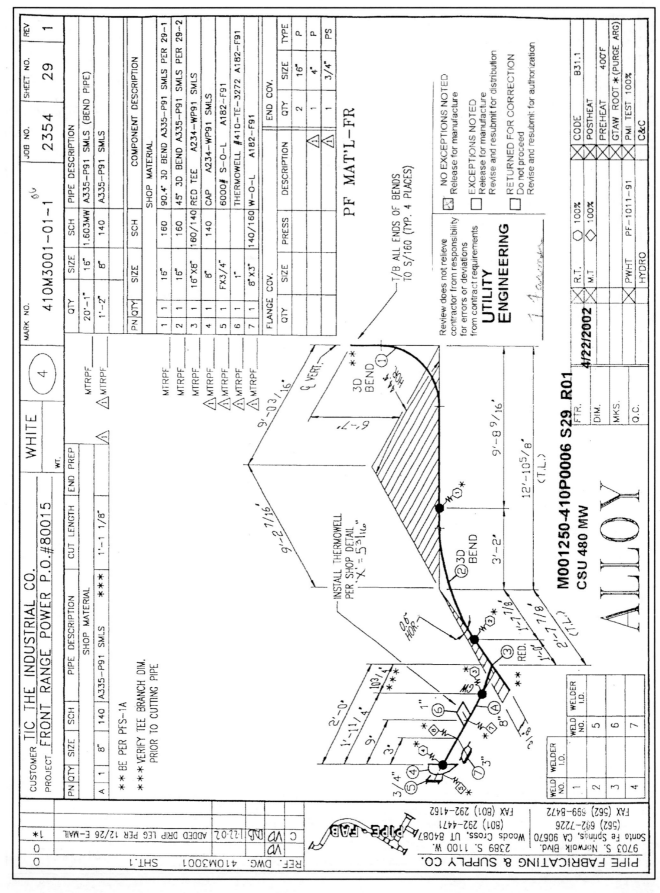

Figure 20 ♦ Spool page 16LKEA-3001 sheet 2.

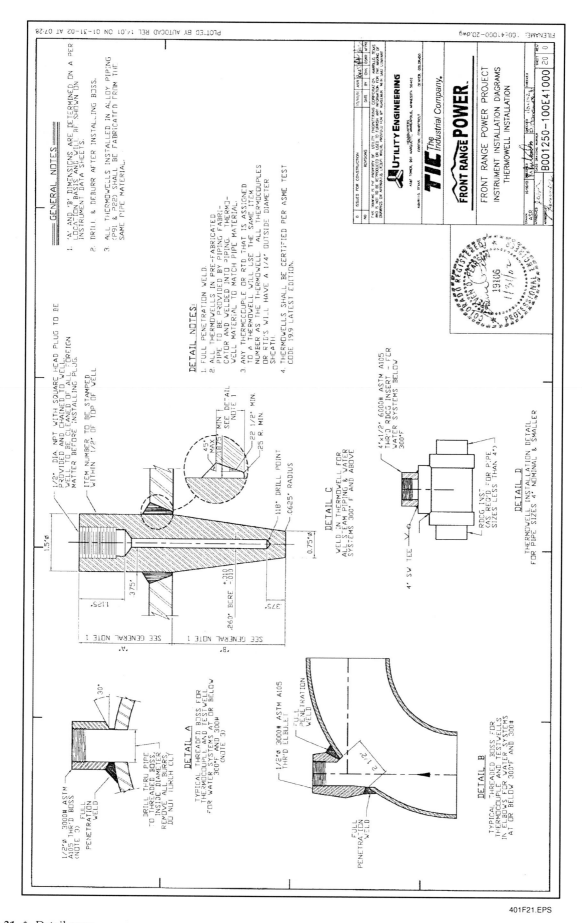

Figure 21 ◆ Detail page.

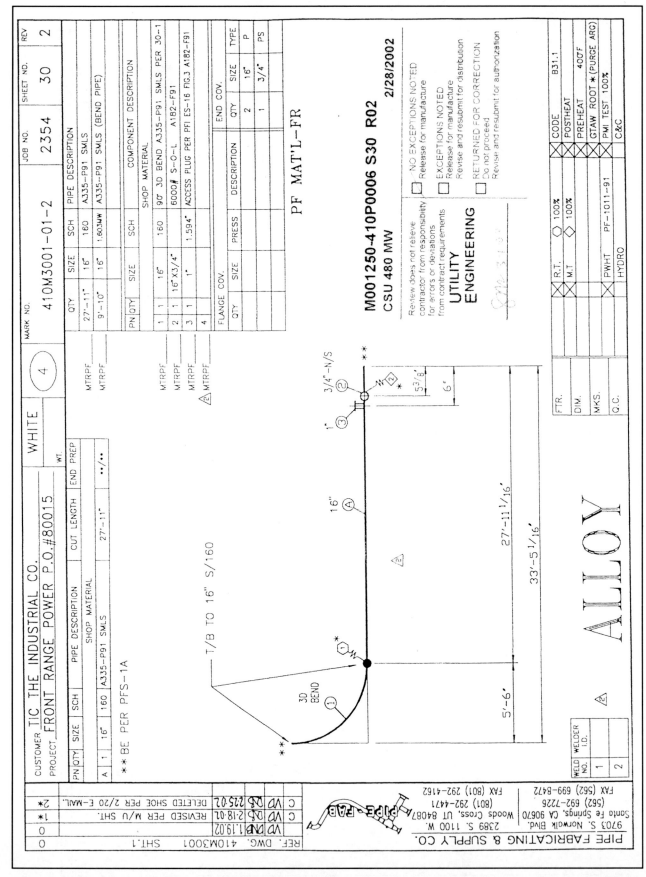

Figure 22 ◆ Spool page 16LKEA-3001 sheet 3.

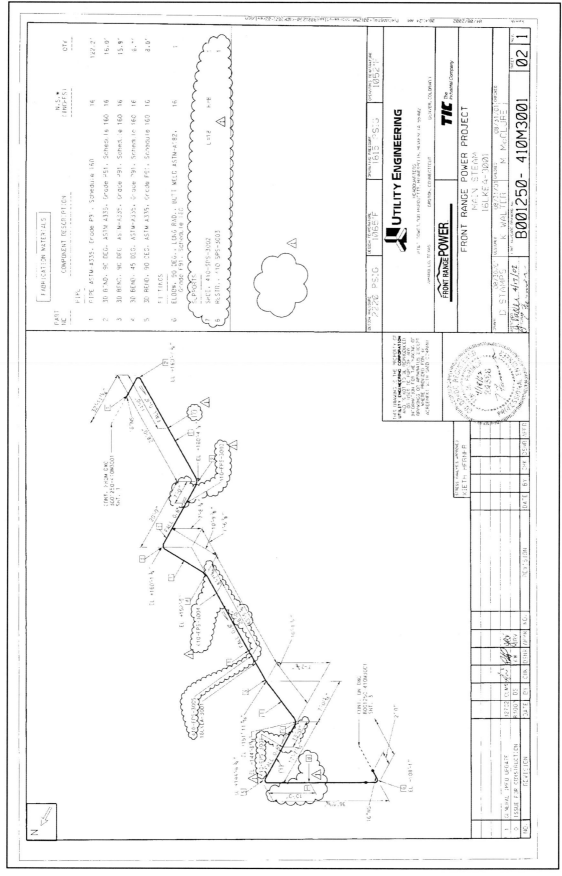

Figure 23 ♦ ISO line 16LKEA-3001 sheet 2.

The spool sheet *(Figure 24)* shows a 1-inch access plug welded on the straight run. This particular plug is at a right angle to the axis of the pipe and to the plane of the bend. On the second spool sheet for this ISO, the detail shows a 1-inch access plug installed at 44.4 degrees from the plane of the bend. The spool sheet that follows *(Figure 26)* shows a 45-degree bend with another 1-inch plug on the far side, as seen in the drawing. The next spool sheet *(Figure 27)* has a single bend that is slightly greater than a 90-degree angle, with 1-inch access plugs 6 inches from each end of the spool. The next spool *(Figure 28)* from the second ISO is marked 410M3001-02-5, and is a straight run with a 45-degree bend. Again, there is a 1-inch access plug on the straight end. The last spool *(Figure 29)* for this ISO is a 90-degree ell on one end and an 89.4-degree bend at a right angle to the plane of the elbow, with a straight joint of pipe 28 feet and $^{11}/_{16}$ inch long in between.

The third ISO *(Figure 30)* matches the single spool sheet for it *(Figure 31)*. There are three ¾-inch sockolets for the pressure instruments, and a 1-inch thermowell, and on the downstream end there is a 1-inch plug.

The last ISO of this line *(Figure 32)* is on two spools. The access plugs for purge are on each end of the first spool *(Figure 33)*. There is a straight and a 90-degree long radius ell with a straight end. The last spool *(Figure 34)* takes the run to the 16 × 14 reducer, which is the end of the pipe contractor's work. There is another 8-inch branch going to a cap and sockolet, with a thermowell and a 3-inch weldolet branch on it. The 16-inch line has a 90-degree ell and then a ¾-inch sockolet horizontally, a ¾-inch sockolet at 45 degrees from vertical, and two 1-inch thermowells at a 45-degree angle from the vertical. Finally, there is the 16 × 14 concentric reducer that ties into the turbine inlet.

7.0.0 ◆ DRAWING ISOs

One of the best ways to understand how ISOs are drawn from plan views is to draw one yourself. The ability to make rough isometric sketches in the field is a useful skill that all pipefitters should develop. ISOs are drawn on special paper that has vertical lines and lines projected at 30 degrees. In this section of the module, you are required to draw a simple ISO from a given plan view. *Figure 35* is a piece from a drawing that shows a plan view of 20 vertical feet of elevation, so it would require an elevation drawing to follow the specific lines.

Find line 221-12LDBA on the plan view.

Before you start drawing the ISO, be sure you understand the plan view fully. Check all dimensions, direction of flow, how far the line drops, its orientation, and how it is put together. Be sure you know all the elevations and whether they are B.O.P. or center line. It may help to make a few rough elevation sketches on scrap paper to help you visualize the line.

Start the ISO by drawing the north arrow first, and be sure you orient the line on the ISO exactly as the plan view shows the line. Make a rough isometric sketch first to make sure you understand how the line should be shown isometrically. Remember, you are not limited by scale.

Pay particular attention to how you dimension the ISO. Put the dimensions in accurately and clearly. Try to avoid cluttering the ISO with all the information you need to show. An effective ISO is pleasing to the eye.

Do not forget match lines, the line number, and the direction of flow. Draw up a complete bill of material for the ISO. Calculate the cut lengths of straight pipe required for each straight section, add them together, and round up to the nearest foot. Be sure to include the following information in the bill of material:

- ISO number
- Line number
- Reference number
- Your name in a box marked Drawn By

Your ISO should be clear enough so that anyone with a knowledge of piping drawings should be able to fabricate and erect the line.

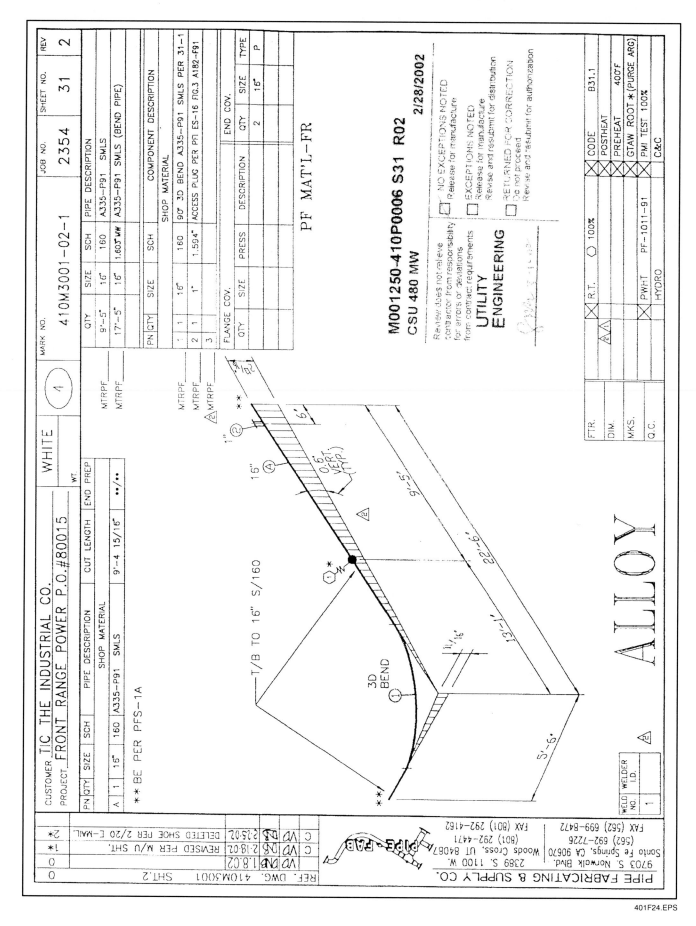

Figure 24 ♦ Spool page 16LKEA-3001 sheet 3.

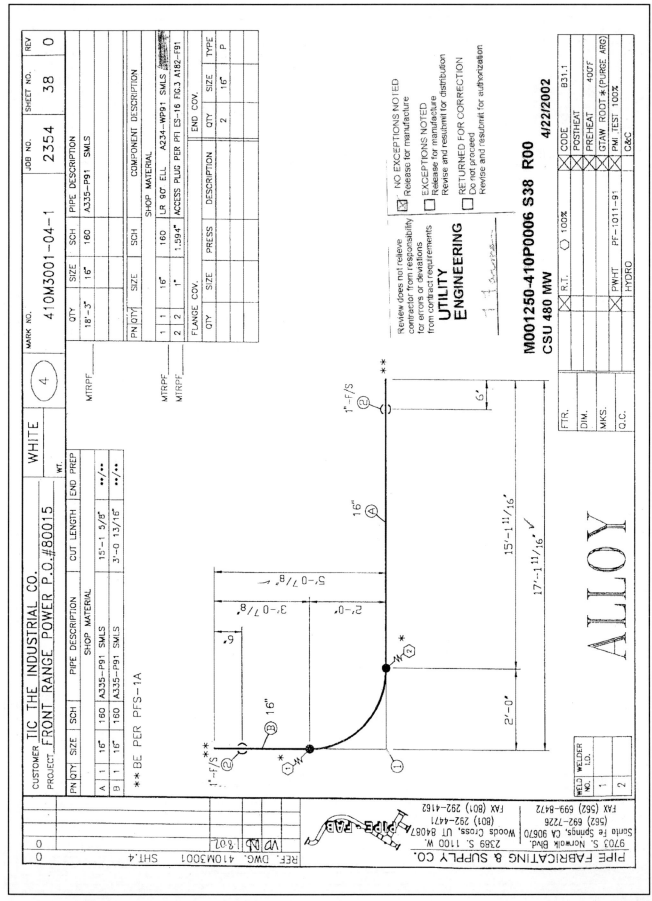

Figure 25 ◆ Spool page 16LKEA-3001 sheet 4.

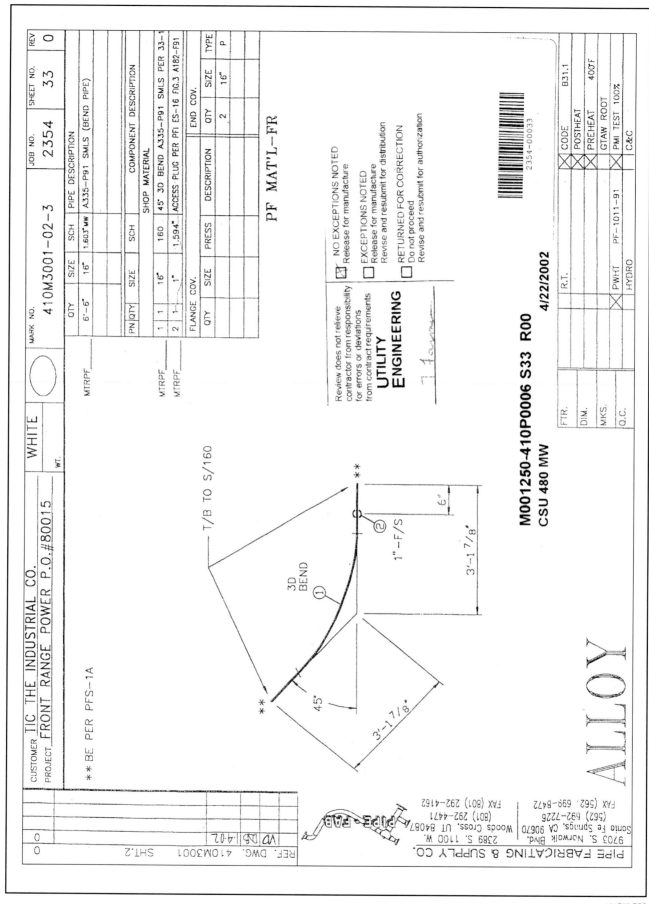

Figure 26 ◆ Spool page 16LKEA-3001 sheet 5.

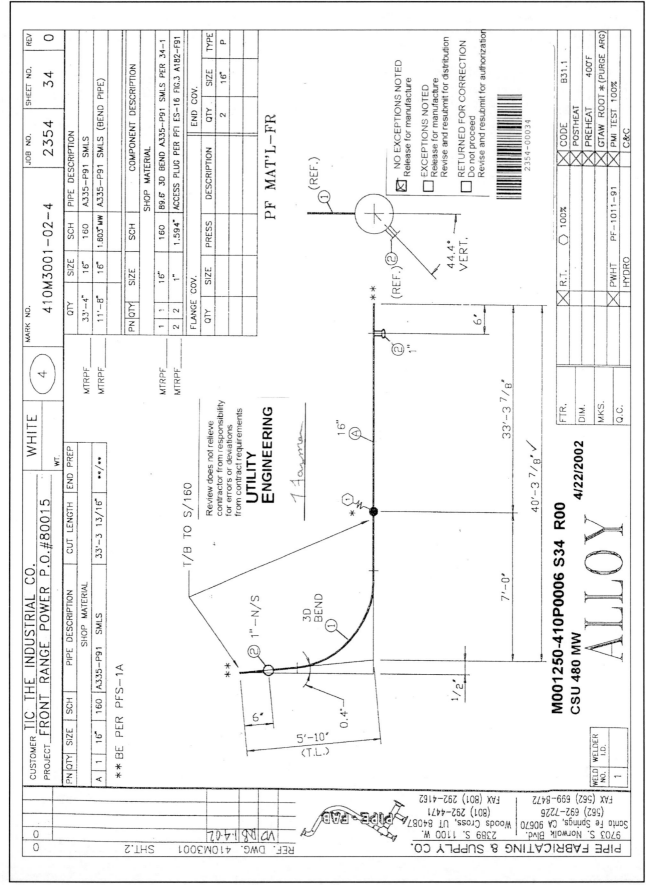

Figure 27 ♦ 3001 sheet 6.

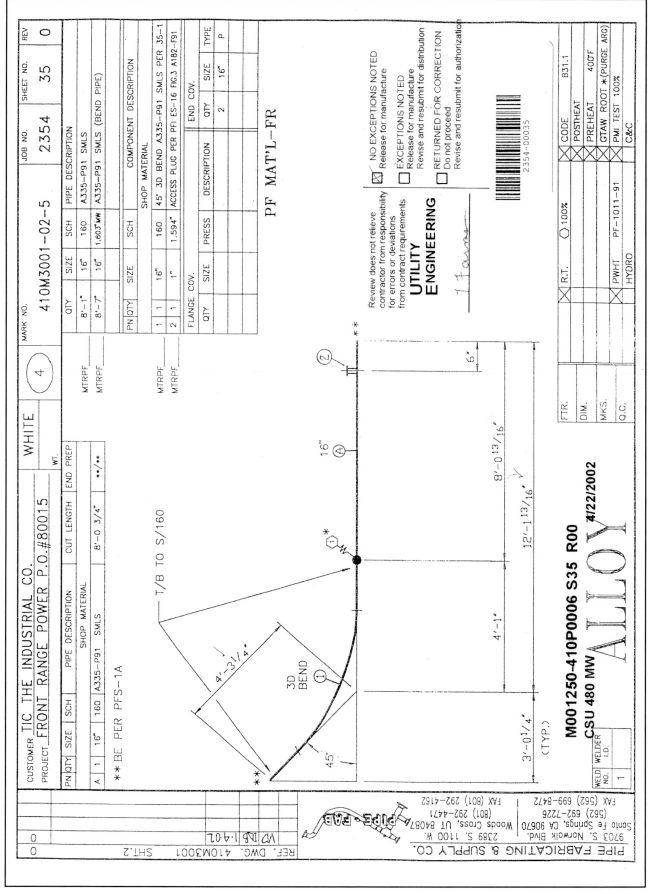

Figure 28 ♦ Spool page 16LKEA-3001 sheet 7.

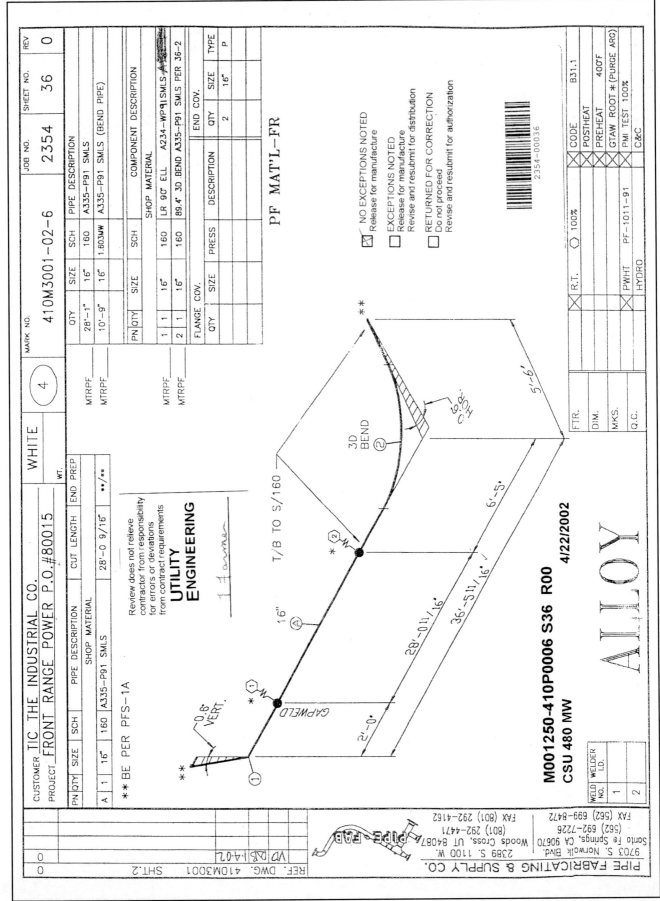

Figure 29 ◆ Spool page 16LKEA-3001 sheet 8.

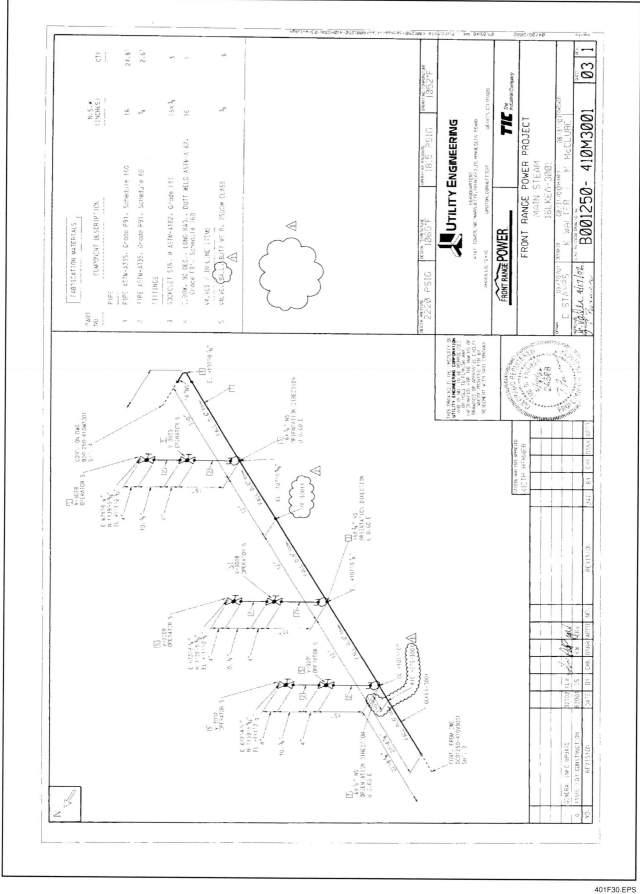

Figure 30 ♦ ISO line 16LKEA-3001 sheet 3.

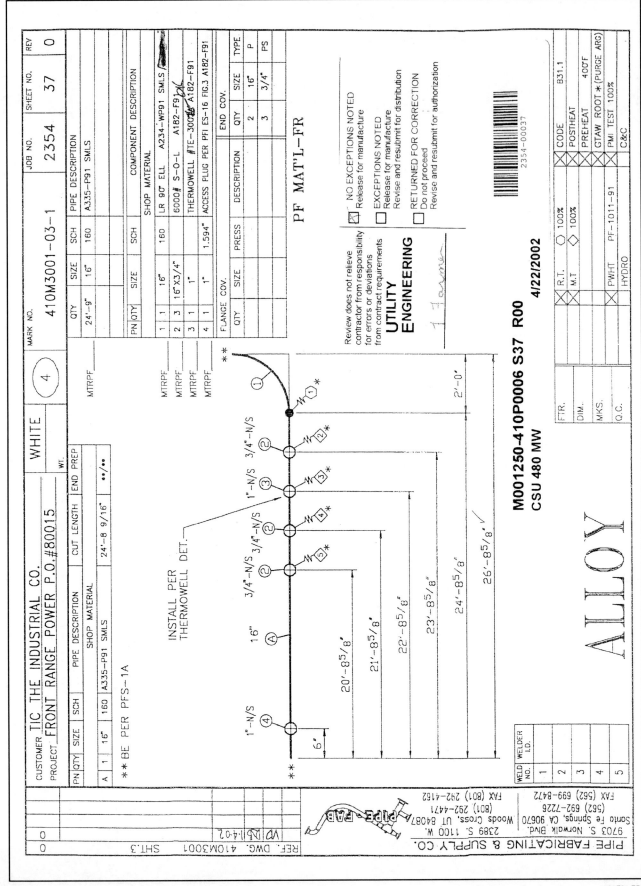

Figure 31 ◆ Spool page 16LKEA-3001 sheet 9.

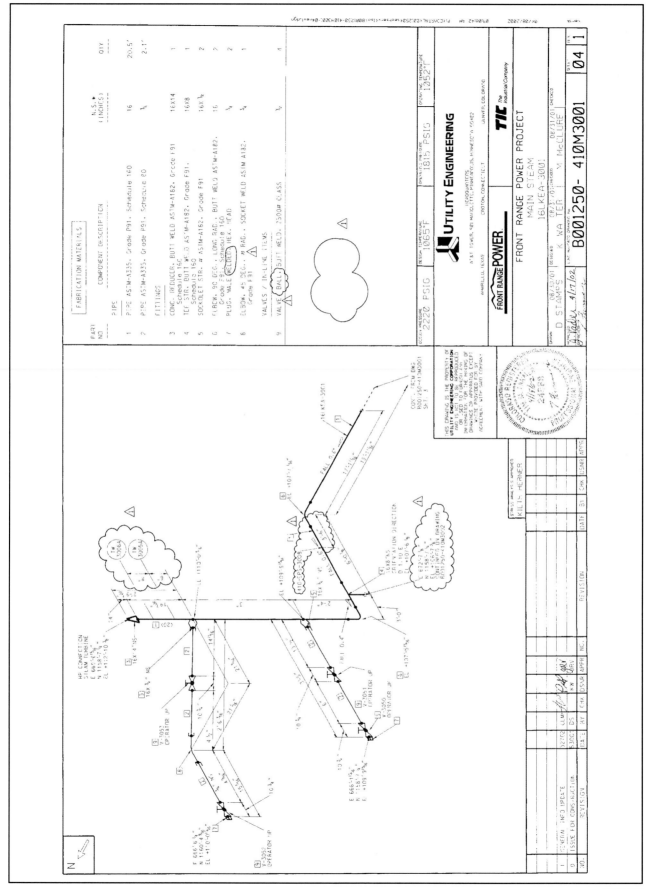

Figure 32 ♦ EA-3001 sheet 4.

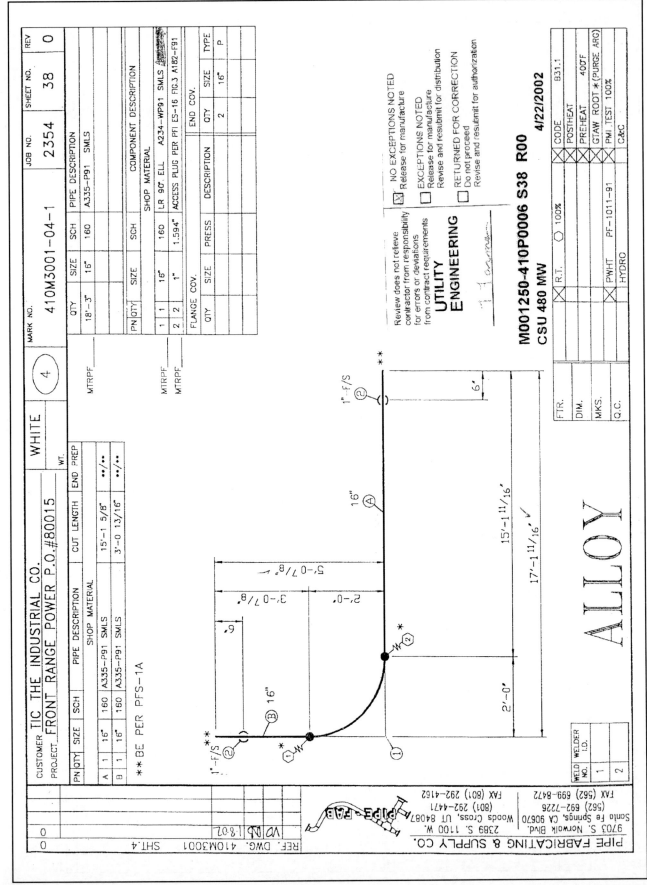

Figure 33 ♦ Spool page 16LKEA-3001 sheet 10.

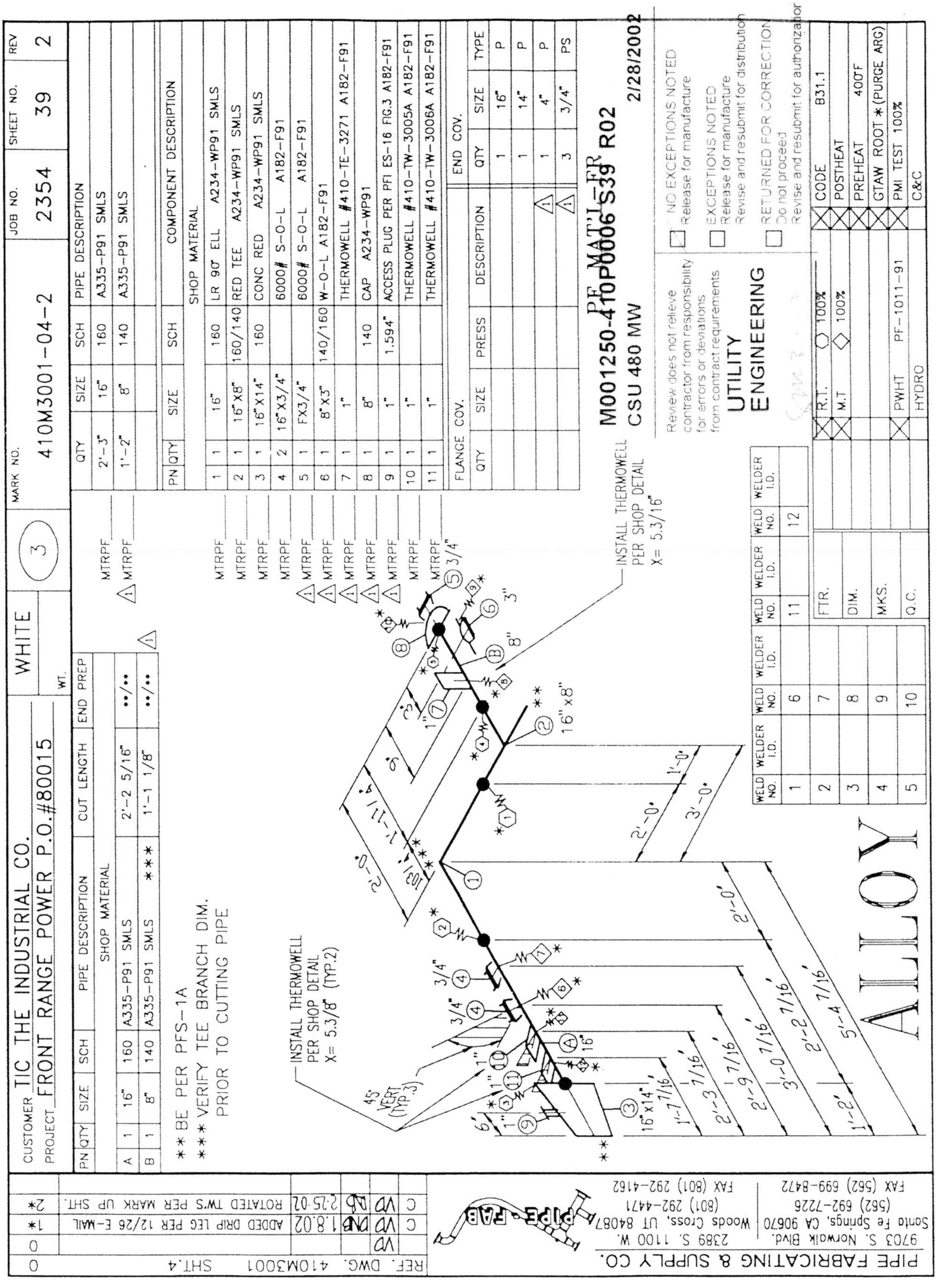

401F34.EPS

Figure 34 ◆ Spool page 16LKEA-3001 sheet 11.

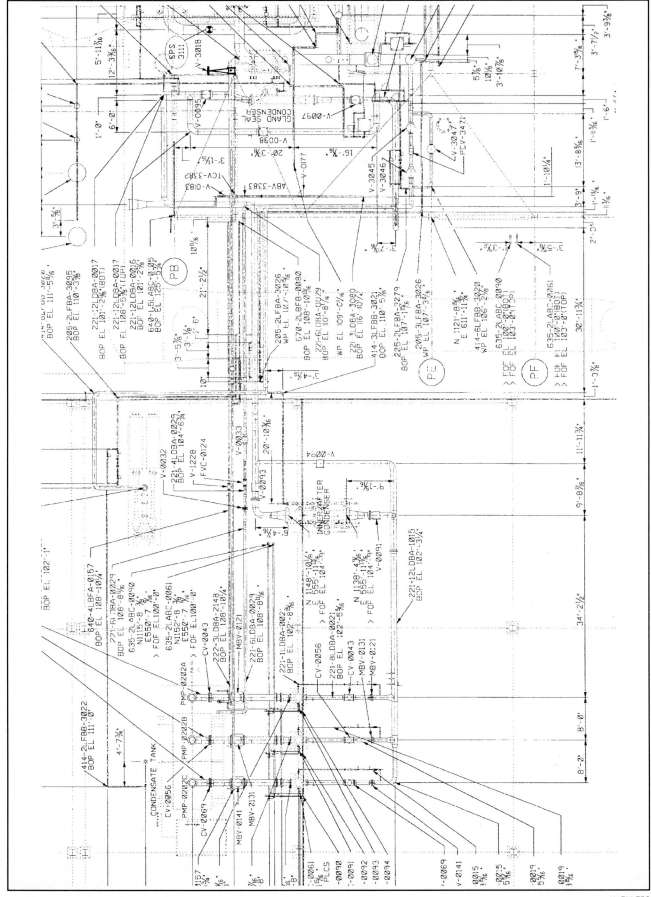

Figure 35 ◆ Piping general arrangement.

Review Questions

1. P&IDs show the _____ of a process.
   a. process flow and function
   b. details
   c. general geographical arrangement
   d. contents

2. The initials HT stand for _____.
   a. hot tap
   b. heat traced
   c. heating terminal
   d. hot water treatment

3. P&IDs provide information about each pipeline through the use of lines, line numbers, and _____.
   a. detail drawings
   b. coordinates
   c. match lines
   d. elevations

4. When a system is shown on more than one sheet of a P&ID, the continuation is shown in a _____.
   a. legend
   b. cloud
   c. note
   d. match line arrow

5. To show the internal arrangement of an object, engineers use a(n) _____.
   a. plan view
   b. section view
   c. elevation
   d. X-ray view

6. Piping arrangement drawings show location of pipelines by using _____.
   a. coordinates, elevations, and control points
   b. P&IDs
   c. spool drawings
   d. plan views

7. Coordinates are a combination of _____.
   a. columns and floors
   b. numbers and letters
   c. pipe tags and labels
   d. words and numbers

Refer to *Figure 1* to answer Questions 8 through 13.

8. The main line on this spool is _____ pipe.
   a. 10-inch
   b. 12-inch
   c. 14-inch
   d. 16-inch

9. There are _____ radiologically tested welds on the spool.
   a. 5
   b. 6
   c. 7
   d. 11

10. The reducing branches are _____ pipe.
    a. stainless steel
    b. 6-inch
    c. 6-inch and 10-inch
    d. 8-inch

11. Part number 4 on the drawing is a _____.
    a. ¾ tee
    b. sensor
    c. weld
    d. 12 × ¾ sockolet

12. Part number 2 on the drawing is a _____.
    a. 1-inch access plug
    b. 1-inch sockolet
    c. 12-inch pipe
    d. 12 × 6 reducing tee

13. Part number 7 is a _____.
    a. 1-inch sockolet
    b. 1-inch thermowell
    c. 6-inch cap
    d. 6-inch tee

Review Questions

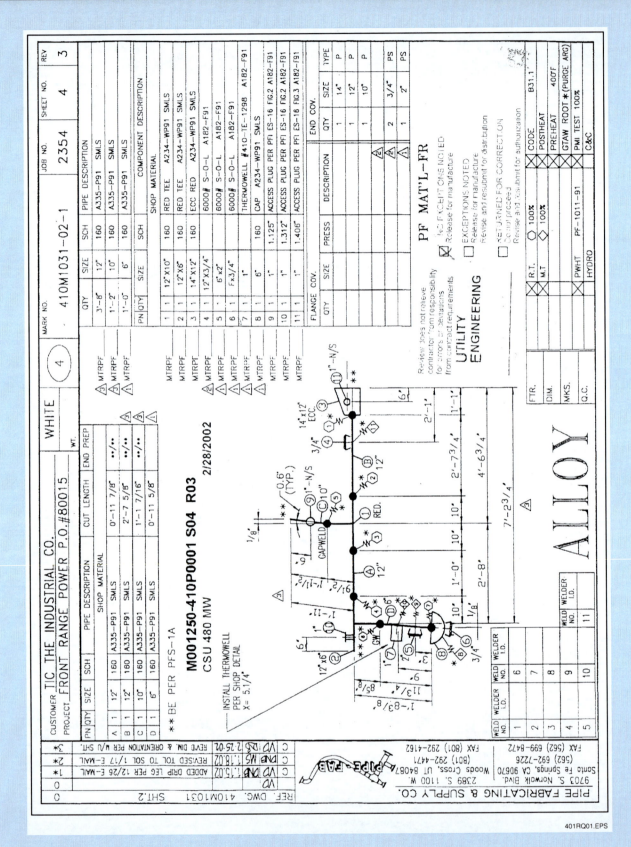

Figure 1

Review Questions

Refer to *Figure 2* to answer Questions 14 through 20.

14. There are _____ feet of 14-inch pipe in this spool.
    a. 7
    b. 37
    c. 38.8
    d. 44

15. Information on the piping run above this is found on drawing _____.
    a. BOO1250-410M3001 Sheet 1
    b. BOO1250-410M2034
    c. BOO1250-410M2031 Sheet 2
    d. BOO1250-410M2032

16. The tee immediately following the eccentric reducer is a _____.
    a. 12 × 6
    b. 12 × 8
    c. 12 × 10
    d. 12 × 12

17. The longest part of the pipe run at this stage is running _____.
    a. due east
    b. a little north of west
    c. a little south of west
    d. a little south of east

18. The valve next to the eccentric reducer is labeled _____.
    a. flat down
    b. 12LKEA-2031
    c. MBV-2001
    d. MBV-2296B

19. The distance from the center of the 14 × 6 tee to the face of valve MBV2001 is _____.
    a. 3'-5"
    b. 4'-6"
    c. 8'-11$\frac{15}{16}$"
    d. 11"

20. The cut length of the spool from the end of MBV2296B to the weld at the 45-degree bend is _____.
    a. 18'-7"
    b. 19'-5"
    c. 20'-5¾"
    d. 24'-10¾"

Review Questions

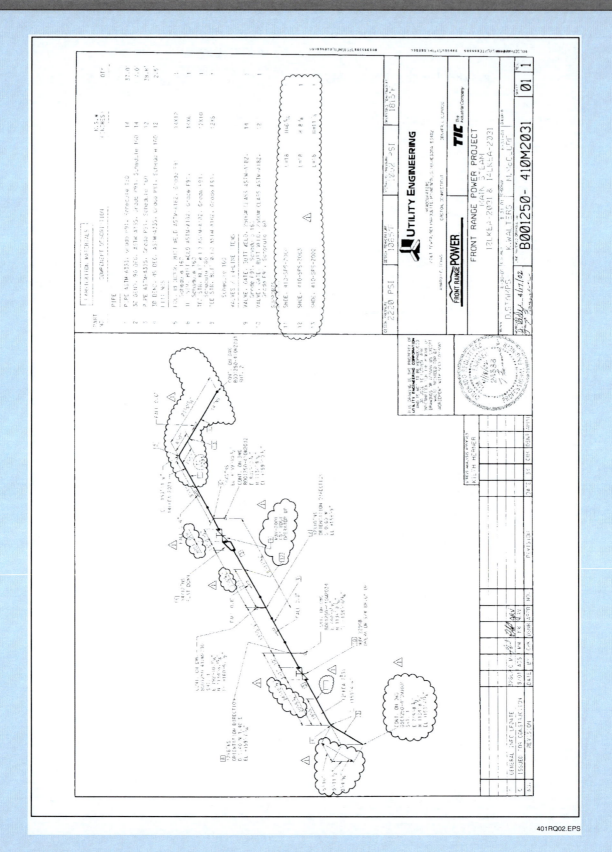

Figure 2

Summary

Blueprint reading is the process by which you as a pipefitter will find your way through a set of blueprints to get the correct pieces together to make a pipeline work. It is important that you examine every relevant sheet for a particular fabrication, to be sure that you adhere to specifications in every way.

Notes

Trade Terms Introduced in This Module

Dimension: A measurement between two points on a drawing.

Revision: A change in a part on an engineering drawing that is noted on the drawing.

Title block: A section of an engineering drawing blocked off for pertinent information, such as the title, drawing number, date, scale, material, draftsperson, and tolerances.

Resources & Acknowledgments

Additional Resources

This module is intended to present thorough resources for task training. The following reference work is suggested for further study. This is optional material for continued education rather than for task training.

Process Piping Drafting, Weaver, Rip; Gulf Publishing Company, Book Division, Houston, TX, 1986.

Figure Credits

Utility Engineering Corp., Blueprints One through Nine; 401F04–401F07, 401F11–401F35, 401RQ01, 401RQ02

NCCER CURRICULA — USER UPDATE

NCCER makes every effort to keep its textbooks up-to-date and free of technical errors. We appreciate your help in this process. If you find an error, a typographical mistake, or an inaccuracy in NCCER's curricula, please fill out this form (or a photocopy), or complete the online form at **www.nccer.org/olf**. Be sure to include the exact module ID number, page number, a detailed description, and your recommended correction. Your input will be brought to the attention of the Authoring Team. Thank you for your assistance.

Instructors – If you have an idea for improving this textbook, or have found that additional materials were necessary to teach this module effectively, please let us know so that we may present your suggestions to the Authoring Team.

NCCER Product Development and Revision
13614 Progress Blvd., Alachua, FL 32615

Email: curriculum@nccer.org
Online: www.nccer.org/olf

❏ Trainee Guide   ❏ AIG   ❏ Exam   ❏ PowerPoints   Other _____

Craft / Level: _____ Copyright Date: _____

Module ID Number / Title: _____

Section Number(s): _____

Description: _____

_____

_____

_____

Recommended Correction: _____

_____

_____

_____

Your Name: _____

Address: _____

Email: _____ Phone: _____

Pipefitting Level Four

08402-07

Advanced Pipe Fabrication

08402-07
Advanced Pipe Fabrication

Topics to be presented in this module include:

1.0.0	Introduction	2.2
2.0.0	Determining Piping Offsets	2.2
3.0.0	Fabricating Miter Turns	2.8
4.0.0	Fishmouth	2.23
5.0.0	Fabricating Dummy Legs and Trunions Out of Pipe	2.30
6.0.0	Laying Out Laterals and Supports Without Using References	2.32

Overview

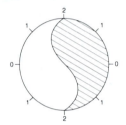

Pipe fabrication is the use of either ordinate tables or trigonometry to fabricate fittings and pipe assemblies to suit a process application. In this module, the trainee learns to use trigonometry to produce ordinates, and to use ordinates to lay out miters and laterals. The trainee also learns the alternative methods of calculating and drawing the ordinates to lay out the cuts for laterals, fishmouths, and mitered turns. The formulas are provided for putting together multiple offsets around an obstacle, either equal spread or unequal spread.

Objectives

When you have completed this module, you will be able to do the following:

1. Calculate simple piping offsets.
2. Calculate three-line, 45-degree, equal-spread offsets around a vessel.
3. Calculate three-line, 45-degree, unequal-spread offsets.
4. Fabricate tank heating coils.
5. Perform mitering procedures.
6. Lay out three- and four-piece mitered turns.
7. Lay out 45-degree laterals, using references or a calculator.
8. Fabricate dummy legs and trunions out of pipe, using references.
9. Perform geometric layout of pipe laterals and supports.
10. Lay out and fabricate a fishmouth.
11. Lay out and fabricate a wye.

Trade Terms

Base line
Branch
Chord
Cutback
Fishmouth
Ordinate
Ordinate lines
Segment

Required Trainee Materials

1. Pencil and paper
2. Appropriate personal protective equipment
3. Scientific calculator
4. Compass
5. 45-degree right triangle
6. Dividers
7. Ruler
8. T-square

Prerequisites

Before you begin this module, it is recommended that you successfully complete *Core Curriculum*; *Pipefitting Level One*; *Pipefitting Level Two*; *Pipefitting Level Three*; and *Pipefitting Level Four*, Module 08401-07.

This course map shows all of the modules in the fourth level of the Pipefitting curriculum. The suggested training order begins at the bottom and proceeds up. Skill levels increase as you advance on the course map. The local Training Program Sponsor may adjust the training order.

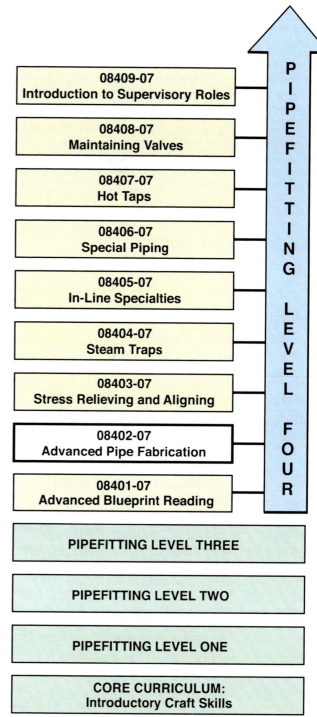

1.0.0 ◆ INTRODUCTION

This module introduces advanced pipe fabrication. Once you are familiar with reading blueprints and fabricating piping components, you are ready to use these skills to perform advanced pipe fabrication activities. Pipefitters must fabricate piping systems that bend around objects, rise or drop to meet other lines, and intersect at angles greater or less than 90 degrees. Although manufactured fittings can be used for most jobs, sometimes their availability or cost prohibit their use. Because of this, the pipefitter must know how to perform advanced pipe fabrication tasks.

2.0.0 ◆ DETERMINING PIPING OFFSETS

An offset is a lateral or vertical move that takes the pipe out of its original line to a line that is parallel with the original line. Offsets are used when it is necessary to change the position of a pipeline in order to avoid an obstruction, such as a wall or a tank.

When determining piping offsets, it is important to know how to solve right triangles. The sum of the three angles of any triangle is 180 degrees. A right triangle has one 90-degree angle and two acute angles, or angles less than 90 degrees, whose sum equals 90 degrees. For example, if one of the acute angles of a right triangle is 45 degrees, the other acute angle is 45 degrees also. If one of the acute angles of a right triangle is 30 degrees, then the other acute angle is 60 degrees. *Figure 1* shows two right triangles.

To solve a right triangle means to find the length of its unknown sides and the degrees of its unknown angles. In order to solve any right triangle you must know the length of two sides or the degrees of two angles, including the 90-degree angle, and the length of one side. *Figure 2* shows a right triangle with two sides known. *Figure 3* shows a right triangle with two angles and one known side.

If these minimums are not known, the triangle cannot be solved. However, in pipe work, this information is given or can be found.

All piping offsets are based on the right triangle. The three basic sides of an offset are the set, run, and travel. The set and run are joined by a 90-degree angle, and the travel connects the span between their end points. *Figure 4* shows the components of an offset.

The set is the distance, measured center-to-center, that the pipeline is offset. The run is the total linear axial distance required for the offset. The travel is the center-to-center measure-

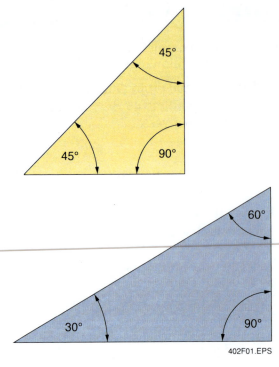

Figure 1 ◆ Right triangles.

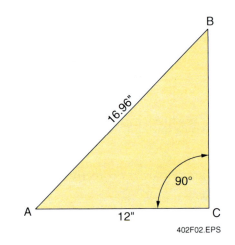

Figure 2 ◆ Right triangle with two sides known.

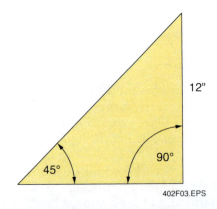

Figure 3 ◆ Right triangle with two angles and one known side.

ment of the offset piping. The angle of the fittings determines the number of degrees that the piping changes direction. The use of right triangle trigonometry will help you calculate the following types of offsets:

- Simple offsets
- Three-line, 45-degree, equal-spread offsets around vessel
- Three-line, 45-degree, unequal-spread offsets
- Tank heating coils

2.1.0 Calculating Simple Offsets

To find the length of the travel for a simple 45-degree offset when the set is known, multiply the set by the constant 1.414. This formula, using the constant of 1.414, only works on a 45-degree offset. Other trig functions are also used to find unknown sides and angles of piping offsets. The most common functions used to calculate simple offsets are sine, cosine, and tangent. *Figure 5* shows the relationship between trig functions and piping offsets.

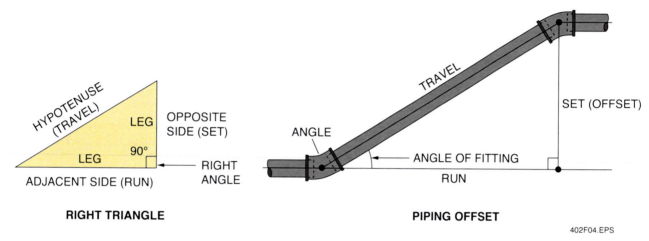

Figure 4 ◆ Offset components.

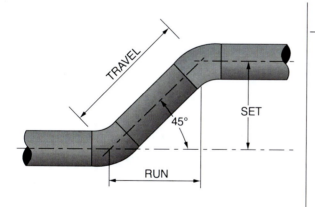

TO DETERMINE ANGLE OF OFFSET WHEN THE LENGTHS OF TWO SIDES ARE KNOWN

SET DIVIDED BY TRAVEL = SINE

RUN DIVIDED BY TRAVEL = COSINE

SET DIVIDED BY RUN = TANGENT

TO DETERMINE LENGTHS OF SIDES WHEN ANGLE AND ONE SIDE ARE KNOWN	ANGLE OF OFFSET				
	60°	45°	30°	22½°	15°
SET = TRAVEL × SINE	0.866	0.707	0.500	0.383	0.259
RUN = TRAVEL × COSINE	0.500	0.707	0.866	0.924	0.966
SET = RUN × TANGENT	1.732	1.000	0.577	0.414	0.268

Figure 5 ◆ Relationship between trig functions and piping offsets.

For this exercise, assume that the set is 57 inches and the pipe is a 4-inch, Schedule 40 butt weld piping system. Follow these steps to find the length of the travel:

Step 1 Multiply the length of the set by 1.414. This gives the length of the travel between the centers of the fittings.

57 × 1.414 = 80.598, or 80⅝ inches

Step 2 Multiply the diameter of the pipe by ⅝ inch (or 0.625) to obtain the takeout for the 45-degree fittings.

4 × 0.625 = 2.5 inches

Step 3 Multiply 2.5 by 2 to obtain the takeout for both 45-degree fittings.

2.5 × 2 = 5 inches

Step 4 Subtract 5 inches from the length of travel measurement from Step 1.

80⅝ − 5 = 75⅝ inches length of travel, less takeout for both 45-degree fittings

Step 5 Subtract ¼ inch from the 75⅝-inch takeout for the two ⅛-inch weld gaps.

75⅝ − ¼ = 75⅜ inches total cut length for the travel of the pipe

NOTE
Different welding procedures will require different weld gaps. For calculation purposes, ⅛ inch is used as an example throughout this module.

2.2.0 Calculating Three-Line, 45-Degree, Equal-Spread Offsets Around a Vessel

Often, you must lay out a series of pipelines in an offset around a vessel and keep the same distance between each line. To do this, you must calculate the distance from the center of the vessel to the first line, the set and travel of that line, and then the offsets of the other lines to be routed around the vessel.

Another way to determine the distance from the center of the vessel to the starting point of the offset is to divide the offset angle by 2 and find the tangent of that angle, which can be found in the *Pipefitters' Handbook* or with a calculator. Multiply the tangent by the distance from the center of the vessel to the center of the pipe. For example: a 45-degree offset divided by 2 equals 22½ degrees, and the tangent for 22½ degrees is 0.41421, or 0.4142. Using 14 inches as the center line of a vessel to the center line of a pipe, the distance from the center of the vessel to the starting point of the 45-degree offset is 14 × 0.4142 = 5.79, or 5¾ inches.

Figure 6 shows a three-line, 45-degree, equal-spread offset.

Refer to *Figure 6* and follow these steps to calculate a three-line, 45-degree, equal-spread offset around a vessel, using 4-inch pipe:

Step 1 Add the distance from the wall to the side of the vessel, plus the diameter of the vessel, plus the distance from the outside of the vessel to the center of line 1 to obtain the distance of line 1 from the wall before the offset around the vessel (measurement A).

3 inches + 18 inches + 5 inches = 26 inches

Step 2 Subtract the distance from the wall to the center of line 1 after the offset around the vessel from the measurement found in Step 1 to obtain the set of line 1.

26 inches − 6 inches = 20 inches

Step 3 Add the radius of the vessel to the distance from the outside of the vessel to the center of line 1, and multiply by 0.4142 to obtain the distance from the center of the vessel to the starting point of the offset for line 1 (measurement B).

9 inches + 5 inches = 14 inches × 0.4142 = 5.79, or 5¹³⁄₁₆ inches

Step 4 Divide the set of line 1 by the 45-degree angle sine, 0.7071, to determine the travel of line 1.

20 inches ÷ 0.7071 = 28.28, or 28¼ inches

NOTE
The set and travel for all three lines in an equal-spread offset is always the same, but the starting points for the three lines will be different.

Step 5 Multiply the distance between lines 1 and 2 by 0.4142 to obtain the starting point of the offset for line 2 (measurement F). This is the distance past the start of the offset of line 1 that the offset of line 2 will start.

9 inches × 0.4142 = 3.727, or 3¾ inches

Step 6 Multiply the distance between line 1 and line 3 by 0.4142 to obtain the starting point of the offset for line 3 (measurement G). This is the distance past the starting point of the offset for line 1 where the offset for line 3 will start.

9 inches + 9 inches × 0.4142 = 7.455, or 7⁷⁄₁₆ inches

2.3.0 Calculating Three-Line, 45-Degree, Unequal-Spread Offsets

Calculating unequal-spread offsets is necessary to move piping around an object. Calculating a three-line, 45-degree, unequal-spread offset is similar to calculating a three-line, 45-degree, equal-spread offset except that the travel lengths will be different. *Figure 7* shows a three-line, 45-degree, unequal-spread offset.

Refer to *Figure 7* and follow these steps to determine the travel length for line 2 (measurement H) and the travel length for line 3 (measurement M) on a three-line, 45-degree, unequal-spread offset:

NOTE
Measurements must start at line 1, and these measurements are used to obtain measurements for lines 2 and 3.

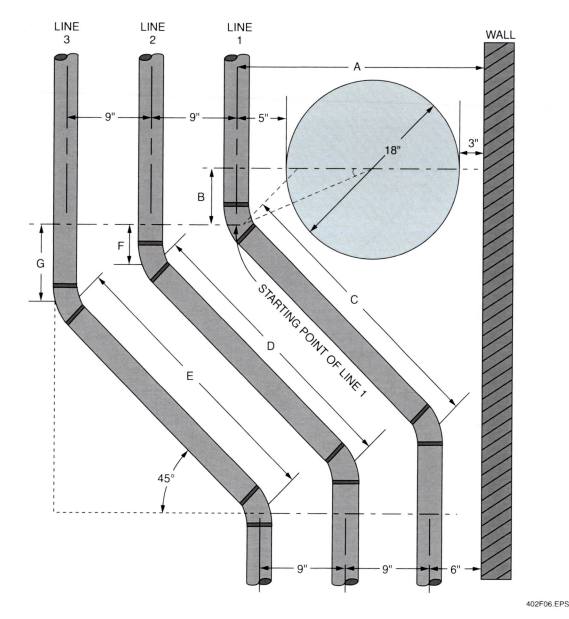

Figure 6 ◆ Three-line, 45-degree, equal-spread offset.

Step 1 Measure the center-to-center distance between line 1 and line 2 at spread 1.

 15 inches

Step 2 Measure the center-to-center distance between line 2 and line 3 at spread 2.

 14 inches

Step 3 Measure the center-to-center distance between line 1 and line 2 at spread 3.

 16 inches

Step 4 Measure the center-to-center distance between line 2 and line 3 at spread 4.

 21 inches

Step 5 Multiply the distance between line 1 and line 2 at spread 1 by 1.414 to obtain the line 1 center line measurement (measurement E).

 15 inches × 1.414 = 21.21 or 21¼"

Step 6 Subtract the distance between line 1 and line 2 at spread 3 from measurement E to obtain the starting point of the line 2 offset (measurement F).

 21.21 inches − 16 inches = 5.21 or 5¼"

Step 7 Multiply the starting point (measurement F) of line 1 by 1.414 to obtain the line 3 center line (measurement G)

 5.21 inches × 1.414 = 7.366 or 7⅜"

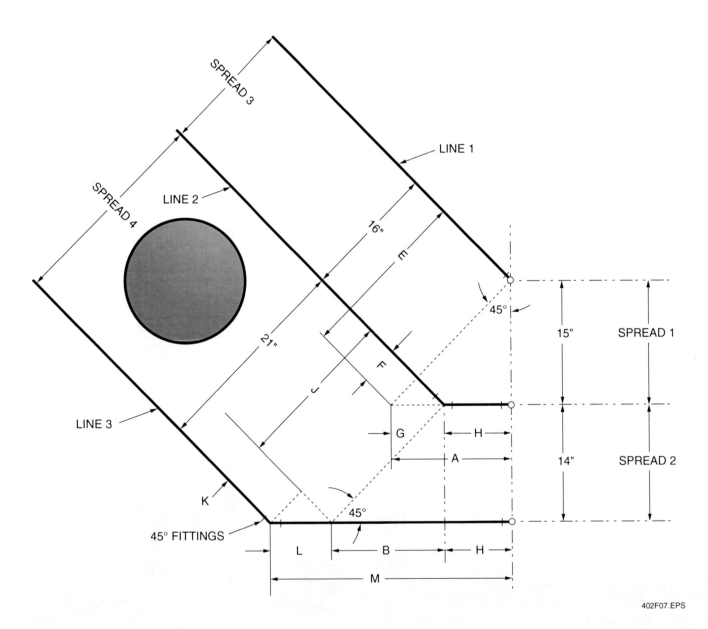

Figure 7 ◆ Three-line, 45-degree, unequal-spread offset.

Step 8 Subtract the starting point of the line 2 offset (measurement G) from the distance between line 1 and line 2 at spread 1 to obtain the travel length (measurement H) for line 2.

15 inches − 7.366 inches = 7.364 or 7⅜"

Step 9 Multiply the distance between line 2 and line 3 at spread 2 by 1.414 to obtain the line 3 center line (measurement J).

14 inches × 1.414 = 19.796 or 9¹³⁄₁₆"

Step 10 Subtract line 2 center line measurement from the center-to-center measurement of line 2 and line 3 at spread 4 to obtain the starting point of the line 3 offset (measurement K).

21 inches − 19.796 = 1.204 or 1³⁄₁₆"

Step 11 Multiply line 3 starting point offset measurement by 1.414 to obtain the intersection of the line 2 offset starting point (measurement L).

1.204 inches × 1.414 = 1.702 or 1¹¹⁄₁₆"

Step 12 Add line 2 travel length plus the center-to-center distance between line 2 and line 3 at spread 2 plus the distance from the start point of line 3 offset to the intersection of line 2 offset to obtain the travel length (measurement M) for line 3.

7.634 inches + 14 inches + 1.702 inches = 23.336, or 23⁵⁄₁₆ inches

Three-line, 45-degree, unequal-spread offsets can be fabricated using different size pipes. For example, line 1 could be a 2-inch pipe, line 2 could be a 3-inch pipe, and line 3 could be a 4-inch pipe. It is important to use the correct center-to-center measurements when calculating the takeouts for the different size pipes.

2.4.0 Laying Out and Fabricating Tank Heating Coils

Although tank coils (*Figure 8*) may be made on a roll bender as a continuous coil, the pipefitter is sometimes called upon to fabricate tank heating coils. A tank coil is a piping system that is fabricated around the inside walls of a tank and keeps the same radius from the center of the tank. A typical tank coil provides steam or heated liquid to a vessel to keep a material molten.

The pipe for a tank coil must be assembled or erected inside the tank. To determine the length of pipe for a tank coil, the pipefitter must know the radius of the tank and the radius of the pipe coil inside the tank. After these measurements are known, a table of constants can be used to obtain the center-to-center measurements of the tank coil. *Table 1* shows a typical table of constants. The number of pieces for each coil is obtained by dividing the fitting angle into 360 degrees.

Refer to *Figure 8* and follow these steps to lay out and fabricate a tank coil:

Step 1 Divide the diameter of the tank by 2 to obtain the radius of the tank (measurement A).

76 inches ÷ 2 = 38 inches

Step 2 Measure the distance from the outside diameter of the tank to the center line of the coil to obtain measurement B.

10 inches

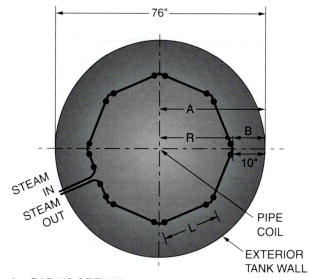

A = RADIUS OF TANK
R = RADIUS OF PIPE COIL

Figure 8 ◆ Tank coil.

Table 1 Table of Constants

Angle of Fitting	Number of Pipes per Coil	Constant
90° (A)	4	1.4142
60° (B)	6	1.0000
45° (A)	8	0.7653
30° (B)	12	0.5176
22½° (B)	16	0.3902
11¼° (B)	32	0.1960
5⅝° (B)	64	0.0981

(A) = Factory fitting
(B) = Fitter will fabricate

Step 3   Subtract the distance from the outside diameter of the tank to the center line of the coil from the radius of the tank to obtain the radius of the coil (measurement R).

     38 inches – 10 inches = 28 inches

Step 4   Multiply the radius of the coil by the constant for a 45-degree fitting to obtain the center-to-center measurement of the pipe travel for the coil (measurement L).

     28 inches × 0.7653 = 21.428,
     or 21 $\frac{7}{16}$ inches

Instead of the table of constants, you can calculate the center-to-center lengths by multiplying 2 times the radius of the coil by the sin of one half the angle on a calculator. The answer will be more accurate, and more easily obtained. In Step 5, if you are fabricating the coil miters, the only addition to the center-to-center necessary would be the ⅛-inch weld gaps.

Step 5   Add the takeout for two 45-degree elbows plus two ⅛-inch weld gaps to obtain the total takeout for seven of the eight pipes in the coil.

     The radius of a 2-inch pipe,
     1¼ inches (45-degree takeout) × 2
     = 2½ inches + ⅛ (weld gap takeout) × 2
     = 2½ + ¼ inch = 2¾ inches total takeout

If you were fabricating miters, this would only be the weld gap.

Step 6   Subtract the takeout from the center-to-center measurement of the pipe travel to obtain the cut length for seven of the eight pipes in the coil.

     21.428 inches – 2.75 inches = 18.67,
     or 18 $\frac{11}{16}$ inches

Step 7   Add the takeout for two 90-degree elbows (3 inches each) and the takeout for two ⅛-inch weld gaps to obtain the total takeout for the supply pipe in the coil. You will also have to subtract 2½ inches (the radius of the two pipes) plus ½ inch to maintain the gap between the inlet and output pipes. Otherwise you would have the two pipes on top of each other.

     3 inches × 2 = 6 inches + ⅛ inch × 2 =
     6¼ inches – 3 inches = 9¼ inches

Step 8   Subtract the takeout from the center-to-center measurement of the pipe travel to obtain the cut length for the supply pipe in the coil.

     21.428 inches – 9.25 inches =
     12.178 inches, or 12 $\frac{3}{16}$ inches

Step 9   Divide the coil supply pipe travel length by 2 to obtain the cut lengths.

     12.178 inches ÷ 2 = 0.6089, or 6 $\frac{1}{16}$ inches

3.0.0 ♦ FABRICATING MITER TURNS

Pipe cannot always run straight. It must be able to bend around objects, rise or drop to meet other lines, and intersect at angles greater or less than 90 degrees. Although manufactured fittings may be used, there may be times when the pipefitter in the field must know how to cut a piece of pipe to a specified angle, called a miter. In order to miter a piece of pipe, you must first learn how to lay out *ordinate lines* and *cutback* lines on a piece of pipe. You will then be ready to fabricate mitered turns.

3.1.0 Laying Out Ordinate Lines

In pipe fabrication, the circumference of a pipe is divided into 4, 8, 16, or more *segments* depending on the size of the pipe and the complexity of the intersection. At each of these division points, a straight line is drawn along the surface of the pipe. These lines are called ordinate lines. Ordinate lines serve as guides for cutting miters and for marking contours. Certain distances are marked off on the ordinate lines to form the angle at which the pipe needs to be cut. Follow these steps to lay out ordinate lines:

Step 1   Obtain a strip of paper with straight edges long enough to wrap around the outside of the pipe 1½ times.

Step 2   Wrap the strip of paper around the outside of the pipe and mark the point at which the strip of paper overlaps (*Figure 9*).

Step 3   Make a square cut on the strip of paper at the mark.

Step 4   Determine the number of *ordinates* needed for the fabrication. Divide ½- to 3-inch pipe into 4 ordinates, 4- to 10-inch pipe into 8 ordinates, and 12-inch pipe and up into 16 ordinates.

Step 5 Fold the paper enough times to obtain the correct number of ordinates. Fold the paper once for 2 ordinates, twice for 4 ordinates, 3 times for 8 ordinates, and 4 times for 16 ordinates. *Figure 10* shows ordinate folds.

Step 6 Cut off a small piece at the corner of each fold, making a notch to provide an ordinate template.

Step 7 Wrap the paper around the pipe squarely.

Step 8 Make a mark on the pipe at each fold on the ordinate template.

Step 9 Draw a straight line along the length of pipe at each of the ordinate marks, as shown in *Figure 11*.

A piece of angle iron is a good device to use to make straight ordinate marks on the outside of the pipe, as shown in *Figure 12*.

3.2.0 Laying Out Cutback Lines

The pipefitter must mark the cutback on the ordinate lines to provide the cut line for the pipe fitting. Cutback refers to a cut made back from a straight line marked around a pipe where the ordinate lines are marked. *The Pipe Fitters Blue Book* and other pipefitting handbooks give the cutback measurements for common miter angles. Always check your handbooks before manually calculating the cutback measurements. Follow these steps to lay out cutback lines:

Step 1 Lay out the correct number of ordinate lines based on the size of pipe being used.

Step 2 Number the ordinate lines, as shown in *Figure 13*.

Step 3 Using your wraparound and soapstone, draw a working line around the pipe at the center of the desired turn. This will give you a reference point for measuring and marking the necessary cutback lines.

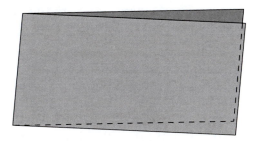

ONE FOLD CREATES
TWO ORDINATES.

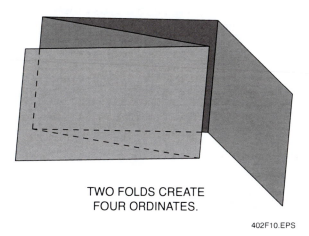

TWO FOLDS CREATE
FOUR ORDINATES.

Figure 10 ◆ Ordinate folds.

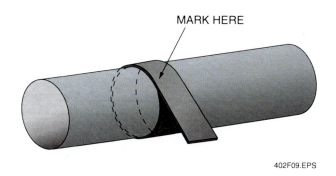

Figure 9 ◆ Strip of paper around a pipe.

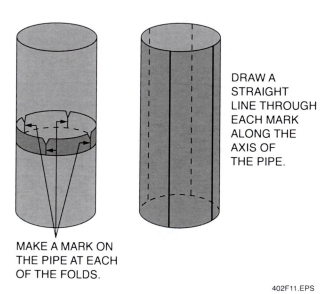

Figure 11 ◆ Ordinate marks.

MODULE 08402-07 ◆ ADVANCED PIPE FABRICATION 2.9

Step 4 Multiply the outside diameter (OD) of the pipe by the tangent of the degree of the miter to be cut and then divide this measurement by 2 to obtain the cutback distance for the given miter.

Step 5 Measure to the left of the working line the cutback distance along ordinate 2 at the top of the pipe and mark the pipe, as shown in *Figure 14*.

Step 6 Measure to the right of the working line the cutback distance along ordinate 2 at the bottom of the pipe and mark the pipe. If more ordinate cutbacks are required, either calculate the cutback lengths by trigonometry or use the given formulas and tables in the pipefitters' handbooks to obtain the cutback measurements. *Figures 15* through *17* show common cutback measurements for various pipe sizes and miter cuts.

There are two methods to determine the ordinate points for a cutback line. The tables in *The Pipe Fitters Blue Book* and other handbooks are one way. The other method uses trigonometry to obtain the ordinates. It is best to know both methods, because the tables don't give every miter angle possible, whereas a trigonometric calculation can. Also, should the pipefitter need to fabricate a very large pipe, trigonometric calculations allow more ordinate points for the longer line involved.

Two calculations are required to obtain the dimensions needed for the layout. First is the length of the initial, longest line from the working line to the place where the miter first strikes the surface of the pipe. That is determined, as you just learned, by multiplying the tangent of the miter angle by the outside diameter of the pipe, and dividing that number by two. This works because the distance from the surface at the working line to the middle of the pipe is one side (the opposite side of the relationship that is defined by the tangent), and the other side (the adjacent side) is the cutback dimension along the surface.

The second calculation, now that the dimension of the first side is given, again uses right angle trigonometry. The two sides that define a right angle are the working line and the axis of the pipe.

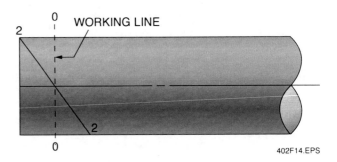

Figure 12 ◆ Using an angle iron to make ordinate marks.

Figure 14 ◆ Conventional method for marking ordinates.

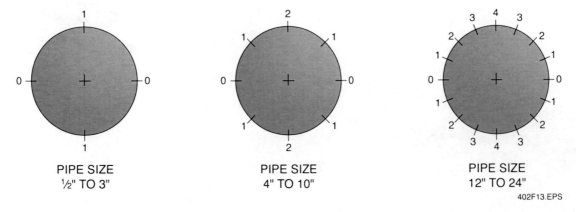

Figure 13 ◆ Numbering ordinates.

2.10 PIPEFITTING ◆ LEVEL FOUR

Divide 90 degrees by the number of points you want, and you will know how far apart in degrees the angle will be from the intersection of the first cutback line to the ends of the other ordinates. The first, or number 1 line in a 4-point quadrant will be 22½ degrees from the vertical working line. Since the straight line from the working line to the ordinate point is the opposite side and the long line to that point is the hypotenuse, the axial line is found by sine 22½ times the axial cutback.

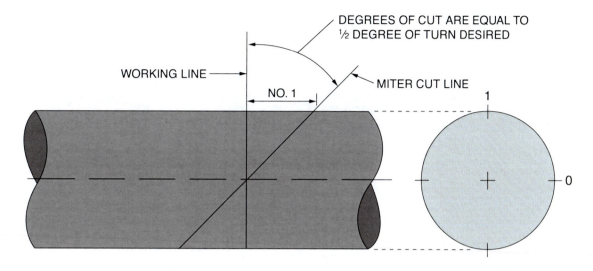

MITER CUTS FOR 1½" THROUGH 3" WITH PIPE DIVIDED INTO 4 ORDINATES.
CUTBACK LINE NO. 1 DIMENSION EQUALS
TANGENT OF CUT × OD OF PIPE DIVIDED BY 2.

1½" THROUGH 3" MITER CUTS – PIPE QUARTERED

7½° CUT FOR 15° TURN		22½° CUT FOR 45° TURN	
SIZE	NO. 1	SIZE	NO. 1
1½	⅛	1½	⅜
2	⅛	2	½
2½	3/16	2½	9/16
3	3/16	3	¾
9° CUT FOR 18° TURN		30° CUT FOR 60° TURN	
SIZE	NO. 1	SIZE	NO. 1
1½	⅛	1½	½
2	3/16	2	11/16
2½	¼	2½	13/16
3	¼	3	1
11¼° CUT FOR 22½° TURN		45° CUT FOR 90° TURN	
SIZE	NO. 1	SIZE	NO. 1
1½	3/16	1½	1 5/16
2	¼	2	1 3/16
2½	¼	2½	1 7/16
3	5/16	3	1 ¾
15° CUT FOR 30° TURN			
SIZE	NO. 1		
1½	¼		
2	5/16		
2½	⅜		
3	7/16		

Figure 15 ♦ Ordinate cutbacks for ½-inch through 3-inch pipe.

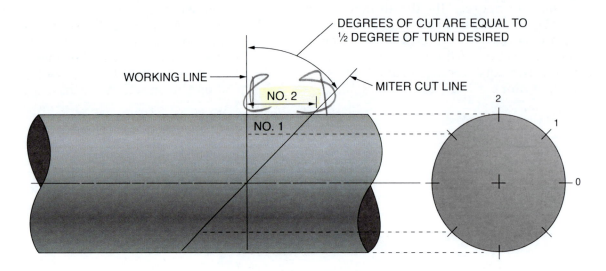

MITER CUTS FOR 4" THROUGH 10" WITH PIPE DIVIDED INTO 8 ORDINATES.
CUTBACK LINE NO. 2 DIMENSION EQUALS
TANGENT OF CUT × OD OF PIPE DIVIDED BY 2.
CUTBACK LINE NO. 1 DIMENSION EQUALS DIMENSION NO. 2 × 0.7071

4" THROUGH 10" MITER CUTS – PIPE IN EIGHTHS

7½° CUT FOR 15° TURN			22½° CUT FOR 45° TURN		
SIZE	NO. 1	NO. 2	SIZE	NO. 1	NO. 2
4	3/16	1/4	4	11/16	15/16
6	5/16	7/16	6	1	1 3/8
8	3/8	9/16	8	1 1/4	1 3/4
10	1/2	11/16	10	1 9/16	2 3/16
9° CUT FOR 18° TURN			30° CUT FOR 60° TURN		
SIZE	NO. 1	NO. 2	SIZE	NO. 1	NO. 2
4	1/4	3/8	4	15/16	1 5/16
6	5/16	1/2	6	1 5/16	1 7/8
8	1/2	11/16	8	1 3/4	2 1/2
10	5/8	7/8	10	2 3/16	3 1/16
11¼° CUT FOR 22½° TURN			45° CUT FOR 90° TURN		
SIZE	NO. 1	NO. 2	SIZE	NO. 1	NO. 2
4	5/16	7/16	4	1 9/16	2 1/4
6	7/16	5/8	6	2 3/8	3 5/16
8	5/8	7/8	8	3 1/16	4 5/16
10	3/4	1 1/16	10	3 13/16	5 3/8
15° CUT FOR 30° TURN					
SIZE	NO. 1	NO. 2			
4	3/8	9/16			
6	5/8	7/8			
8	13/16	1 1/8			
10	1	1 7/16			

Figure 16 ♦ Ordinate cutbacks for 4-inch through 10-inch pipe.

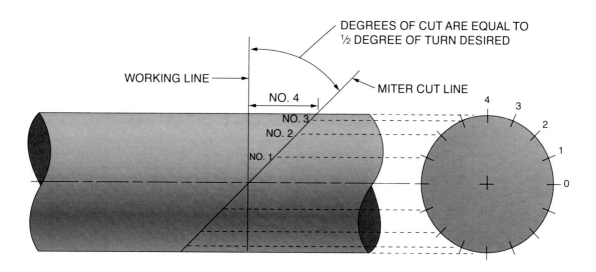

MITER CUTS FOR 12" THROUGH 24" WITH PIPE DIVIDED INTO 16 ORDINATES.
CUTBACK LINE NO. 4 DIMENSION EQUALS
TANGENT OF CUT × OD OF PIPE DIVIDED BY 2.
CUTBACK LINE NO. 3 DIMENSION EQUALS DIMENSION NO. 4 × 0.9239
CUTBACK LINE NO. 2 DIMENSION EQUALS DIMENSION NO. 4 × 0.7071
CUTBACK LINE NO. 1 DIMENSION EQUALS DIMENSION NO. 4 × 0.3827

12" THROUGH 24" MITER CUTS MARK PIPE IN SIXTEENTHS

7½° CUT FOR 15° TURN				
SIZE	NO. 1	NO. 2	NO. 3	NO. 4
12	5/16	9/16	3/4	13/16
14	3/8	5/8	7/8	15/16
16	7/16	3/4	1	1 1/16
18	7/16	13/16	1 1/16	1 3/16
20	1/2	15/16	1 3/16	1 5/16
24	5/8	1 1/8	1 7/16	1 9/16

9° CUT FOR 18° TURN				
SIZE	NO. 1	NO. 2	NO. 3	NO. 4
12	3/8	11/16	15/16	1
14	7/16	13/16	1	1 1/8
16	1/2	7/8	1 3/16	1 1/4
18	9/16	1	1 5/16	1 7/16
20	5/8	1/8	1 7/16	1 9/16
24	3/4	1 5/16	1 3/4	1 7/8

11½° CUT FOR 22½° TURN				
SIZE	NO. 1	NO. 2	NO. 3	NO. 4
12	1/2	7/8	1 3/16	1 1/4
14	1/2	1	1 5/16	1 3/8
16	5/8	1 1/8	1 7/16	1 9/16
18	11/16	1 1/4	1 11/16	1 12/16
20	3/4	1 3/8	1 13/16	2
24	15/16	1 11/16	2 3/16	2 3/8

12" THROUGH 24" MITER CUTS MARK PIPE IN SIXTEENTHS

15° CUT FOR 30° TURN				
SIZE	NO. 1	NO. 2	NO. 3	NO. 4
12	5/8	1 3/16	1 9/16	1 11/16
14	3/4	1 5/16	1 3/4	1 7/8
16	3/16	1 1/2	2	2 1/8
18	15/16	1 11/16	2 1/4	2 3/8
20	1	1 7/8	2 1/2	2 11/16
24	1 1/4	2 1/4	3	3 3/16

22½° CUT FOR 45° TURN				
SIZE	NO. 1	NO. 2	NO. 3	NO. 4
12	1	1 7/8	2 7/16	2 5/8
14	1 1/8	2 1/16	2 11/16	2 7/8
16	1 1/4	2 5/16	3 1/16	3 5/16
18	1 7/16	2 5/8	3 7/16	3 3/4
20	1 9/16	2 15/16	3 13/16	4 1/8
24	1 7/8	3 1/2	4 5/8	5

30° CUT FOR 60° TURN				
SIZE	NO. 1	NO. 2	NO. 3	NO. 4
12	1 3/8	2 5/8	3 3/8	3 11/16
14	1 9/16	2 7/8	3 3/4	4 1/16
16	1 3/4	3 1/4	4 1/4	4 5/8
18	2	3 11/16	4 13/16	5 3/16
20	2 3/16	4 1/16	5 5/16	5 3/4
24	2 5/8	4 7/8	6 3/8	6 15/16

Figure 17 ♦ Ordinate cutbacks for 12-inch through 24-inch pipe.

Each line thereafter adds 22½ degrees to the calculation. The second line is sine 45 degrees times the longest line; the third line is the sine of 67½ degrees times the longest line.

Had this been a large diameter pipe, such as a 36- or 48-inch pipe, you might have wanted to increase the accuracy of the cutline layout by laying out more ordinates. In such a case, you might have chosen to divide the quadrant into six equal parts, and divide the 90 degrees by six, giving the multipliers as sine 15 degrees, sine 30 degrees, sine 45 degrees, sine 60 degrees, and sine 75 degrees. The procedure is the same.

Step 7 Line up a wraparound on the pipe at the ordinate cutback marks on the right side of the working line (see *Figure 18*).

> **NOTE**
> Remember that you will be working on the right side of the working line and then the left side of the working line.

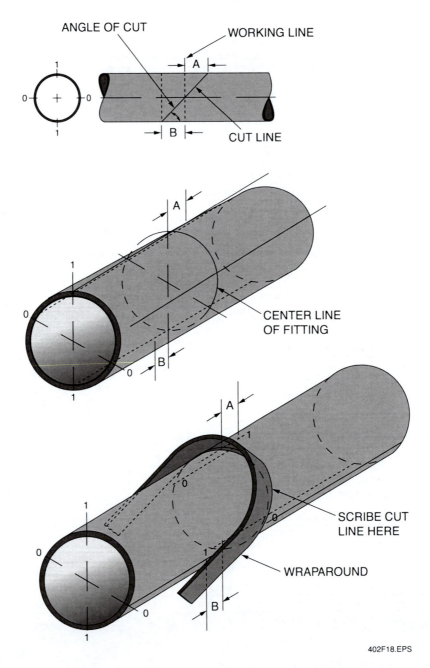

Figure 18 ◆ Laying out cutback marks with wraparound.

Step 8 Connect these marks using soapstone.

Step 9 Line up a wraparound on the pipe at the ordinate cutback marks on the left side of the working line.

Step 10 Connect these marks using soapstone.

Step 11 Secure the pipe in a vise and cut along the line, using a cutting torch. It is important to rotate the pipe to ensure that the torch angle is always at a 90-degree angle with the pipe.

Step 12 Check the miter cut for the correct angle.

Step 13 Bevel the edges of the cut after the cut has been made.

3.2.1 Mitering Exercise

You are now ready to practice cutting a miter using the procedures previously described. Follow these steps to cut a 22½-degree miter cut on 6-inch pipe:

Step 1 Support a 6-inch pipe in two jack stands.

Step 2 Lay out eight ordinate lines on the pipe.

Step 3 Use a wraparound to mark a working line in the center of the desired turn.

Step 4 Find the dimension of cutback line number 2 by using the given formula or checking in the pipefitters' handbook. For a 22½-degree miter on 6-inch pipe, your calculations should be as follows: pipe OD times tangent of 22½-degrees divided by 2, or (6.625 × 0.4142) ÷ 2, or 2.744 ÷ 2, which equals 1.372 or 1⅜ inches. You can also obtain this information from a pipefitting handbook.

Step 5 Measure and mark the cutback along the number 2 ordinate lines and to the left and right side of the working line.

Step 6 Find the dimension for ordinate number 1 by using the given formula or checking in the pipefitters' handbook. For a 22½-degree miter on 6-inch pipe, your calculations should be as follows: dimension of ordinate number 2 times 0.7071, or 1.372 × 0.7071 which equals 0.9701 inches, rounded up to 1 inch.

Step 7 Measure and mark the cutback of 1 inch along the number 1 ordinate lines to the left and to the right of the working line.

Step 8 Use a wraparound to connect the points on the number 0, 1, and 2 cutback marks on the right side of the working line.

Step 9 Use a wraparound to connect the points on the number 0, 1, and 2 cutback marks on the left side of the working line.

Step 10 Secure the pipe in a vise and cut along the line using a cutting torch. It is important to rotate the pipe to ensure that the torch angle is always at a 90-degree angle with the pipe.

Step 11 Check the miter cut for the correct angle.

Step 12 Bevel the edges of the cut after the cut has been made.

3.3.0 Laying Out Mitered Turns

When building a miter, the first consideration should be the radius. The normal rule of thumb is to build the miter similar to the same radius as a standard long radius 90-degree elbow for the size of pipe you are using. A mitered turn (*Figure 19*) consists of several mitered pieces of pipe assembled so that they form a turn of a specific number of degrees with a specific radius. Each mitered piece of pipe is known as a segment. Normally, the number of segments used in 90-degree mitered turns is as follows:

- *Pipe smaller than 6 inches* – 2 to 3 segments
- *Pipe 6 to 10 inches* – 3 to 4 segments
- *Pipe 12 inches and over* – 4 to 7 segments

There are other factors used to determine the number of segments to use in a mitered turn, such as the radius of the turn, engineering specifications, and flow restrictions. The fewer the segments in a mitered turn, the greater the restriction to flow. In order to fabricate a mitered turn, you must either know or calculate the following:

- OD of pipe
- Radius of turn
- Number of degrees in each miter cut
- Center-to-center length of each piece in turn
- Length of end pieces for given radius

In most cases, the radius of the turn is given. When it is not, it can be measured in the field at the location of the pipe run.

To find the number of degrees in each miter cut of a turn, use the following formula:

$$\frac{\text{Number of degrees in turn}}{2 \times \text{number of welds}}$$

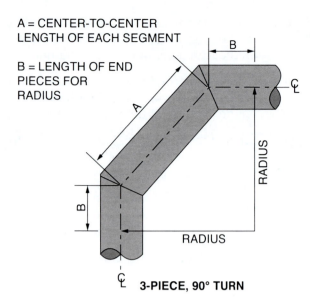

A = CENTER-TO-CENTER LENGTH OF EACH SEGMENT
B = LENGTH OF END PIECES FOR RADIUS

3-PIECE, 90° TURN

$$\text{DEGREES IN EACH MITER} = \frac{\text{NUMBER OF DEGREES IN TURN}}{2 \times \text{NUMBER OF WELDS}}$$

LENGTH OF A = 2 × RADIUS × TANGENT OF MITER
LENGTH OF B = ½ (LENGTH OF A)

Figure 19 ◆ Mitered turn.

The reason the number of welds is used in this formula is that the number of welds is always the total number of pieces in the turn minus 1.

Example: Calculate the number of degrees per miter in a three-piece, 90-degree elbow (two welds).

90 degrees ÷ (2 × 2) = 22½ degrees per miter

3.3.1 Finding Center-to-Center Length

In order for there to be a smooth transition between the segments in a mitered elbow, two things must be consistent: the number of degrees in each miter and the center-to-center length of each segment.

To find the center-to-center length of each segment in a mitered turn, use the following formula:

A = 2 × radius × tangent of miter

Example: The center-to-center length of a three-piece, mitered, 90-degree elbow with a 12-inch radius is found as follows:

A = 2 × radius × tangent of miter
A = 2 × 12 × 0.4142
A = 24 × 0.4142
A = 9.94, or 9¹⁵⁄₁₆ inches

3.3.2 Finding Length of End Pieces

The length of the end pieces for a given radius is calculated by multiplying the center-to-center length by one-half. Therefore, the formula is as follows:

B = A × ½

Using the previous example of a three-piece, mitered, 90-degree elbow on a 12-inch radius with an A dimension of 9¹⁵⁄₁₆ inches, dimension B is calculated as follows:

B = A × ½
B = 9¹⁵⁄₁₆ × ½
B = approximately 5 inches

The following sections explain how to lay out and fabricate three- and four-piece, mitered, 90-degree turns. Lay out the complete turn on a straight length of pipe before cutting out the segments. This saves time because all the measuring and cutting can be done at once.

THREE PIECE 90° TURN
TWO 45° TURNS EQUALS 22½° CUTS

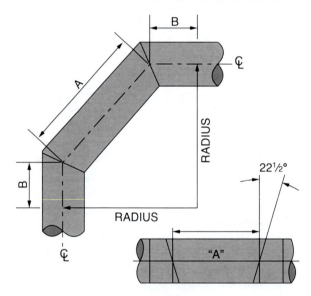

LENGTH "A" EQUALS RADIUS × 0.4142 × 2

RADIUS (INCHES)	LENGTH "A" (INCHES)
12	9¹⁵⁄₁₆
18	14⅞
24	19¾
30	24⅞
36	29⅞
42	34¾
48	39¾

Figure 20 ◆ Three-piece, 90-degree turn.

3.4.0 Laying Out and Fabricating Three-Piece, 90-Degree Mitered Turns

For training purposes, assume that the task is to lay out and fabricate a three-piece, mitered, 90-degree turn on a 12-inch radius, using 8-inch pipe. *Figure 20* shows the completed turn.

Follow these steps to lay out and fabricate a three-piece, mitered, 90-degree turn:

Step 1 Find the number of degrees in each miter.

$$\frac{\text{Number of degrees in turn}}{2 \times \text{number of welds}}$$

$$= \frac{90 \text{ degrees}}{2 \times 2}$$

$$= 22\frac{1}{2} \text{ degrees}$$

Step 2 Find the cutback for a 22½-degree miter on an 8-inch, Schedule 40 pipe.

Cutback = pipe OD × tangent of miter/2

Cutback = (8.625 × 0.4142)/2

Cutback = 3\%6 inches/2

Cutback = 1.786295988 or 1^{13}⁄$_{16}$ inches

Step 3 Find dimension A for a 12-inch radius.

A = 2 × radius × tangent of miter

A = 2 × 12 × 0.41421

A = 24 × 0.41421

A = 9.94, or 9^{15}⁄$_{16}$ inches

Step 4 Divide dimension A in half to find dimension B.

B = 9^{15}⁄$_{16}$ × ½

B = approximately 5 inches

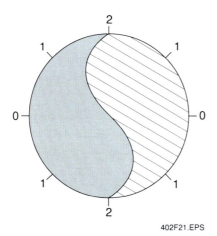

Figure 21 ◆ Numbering ordinates.

Step 5 Lay out eight ordinates on the pipe, using an angle iron as a straight edge, and number the ordinates, as shown in *Figure 21*.

Step 6 Establish a working line or **base line** about 3 inches from the end of the pipe using a wraparound.

Step 7 Measure off dimension B (5 inches) from this working line along ordinates 0 and 0.

Step 8 Draw a line around the pipe connecting the measurement at ordinates 0 and 0, using a wraparound, to establish a working line to lay out dimension B.

Step 9 Measure off dimension A (9^{15}⁄$_{16}$ inches) from this working line along ordinates 0 and 0 and draw a line around the pipe at this point. This line becomes the working line used to lay out dimension A.

Step 10 Measure off dimension B (5 inches) from this working line along ordinates 0 and 0.

Step 11 Draw a line around the pipe connecting these points using a wraparound. This line becomes the working line used to lay out dimension B.

Step 12 Lay out the cutback lines on each of the working lines for a 22½-degree miter. Remember that each miter slants in the opposite direction from the previous one. For 8-inch pipe being mitered 22½-degrees and divided into 8 ordinates, ordinate number 1 will be 1¼ inches from the working line and ordinate number 2 will be 1¾ inches from the working line. *Figure 22* shows the layout of a three-piece, mitered turn.

Step 13 Center-punch marks on either side of ordinates 2 and 2. This helps line up the segments during fit-up.

Step 14 Cut out each segment carefully using an oxyacetylene torch.

Step 15 Dress the edges using a sander or grinder to obtain the proper fit-up.

Step 16 Align the center punch marks and ordinate lines of two segments, and place spacer wires between them. Remember that a certain amount of draw will occur when the segments are tack-welded.

Step 17 Have a qualified welder tack-weld the joint. Adjust the mitered turn as necessary.

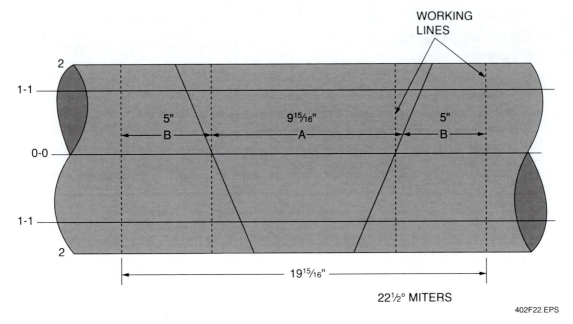

Figure 22 ◆ Layout of three-piece, mitered turn.

Step 18 Align the remaining segment and have a qualified welder tack-weld the joint. Adjust the mitered turn as necessary.

Step 19 Measure the radius to check the squareness of the turn, as shown in *Figure 23*.

3.5.0 Laying Out and Fabricating Four-Piece, 90-Degree Mitered Turns

In this task, you are required to lay out and fabricate a four-piece, mitered, 90-degree turn on an 18-inch radius, using 12-inch pipe. *Figure 24* shows a finished four-piece, 90-degree turn.

Follow these steps to lay out and fabricate a four-piece, 90-degree, mitered turn:

Step 1 Find the number of degrees in each miter.

$$\frac{\text{Number of degrees in turn}}{2 \times \text{number of welds}}$$

$$= \frac{90 \text{ degrees}}{2 \times 3}$$

$$= 15 \text{ degrees}$$

Step 2 Find the cutback for a 15-degree miter on a 12-inch, standard-weight pipe.

Cutback = pipe OD × tangent of miter
Cutback = (12.75 × 0.2679)/2
Cutback = 3.415/2
Cutback = 1.7075 or approximately 1¹¹⁄₁₆ inches

Step 3 Find dimension A for an 18-inch radius.

A = 2 × radius × tangent of miter
A = 2 × 18 × 0.2679
A = 9.64, or approximately 9⅝ inches

Step 4 Divide dimension A in half to find dimension B.

B = 9⅝ × ½
B = 4¹³⁄₁₆ inches

Step 5 Lay out and number 16 ordinates on the pipe.

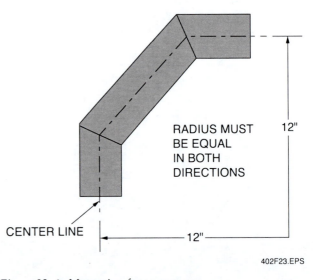

Figure 23 ◆ Measuring for squareness.

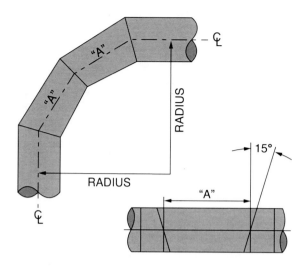

**FOUR PIECE 90° TURN
THREE 30° TURNS WITH 15° CUTS**

LENGTH "A" EQUALS RADIUS × 0.2679 × 2

RADIUS (INCHES)	LENGTH "A" (INCHES)
24	12⅞
30	16 1/16
36	19 5/16
42	22½
48	25¾
60	32⅛
72	38 9/16
84	45
96	51 7/16

Figure 24 ♦ Four-piece, 90-degree turn.

Step 6 Establish a working line about 3 inches from the end of the pipe using a wraparound.

Step 7 Measure off dimension B (4 13/16 inches) from this working line along ordinates 0 and 0.

Step 8 Draw a line around the pipe along dimension B, using a wraparound, to establish a new working line.

Step 9 Measure off dimension A (9⅝ inches) from this working line along ordinates 0 and 0.

Step 10 Draw a line at this point, using a wraparound, to establish a new working line.

Step 11 Measure off another dimension A (9⅝ inches) from this working line along ordinates 0 and 0.

Step 12 Draw a line at this point, using a wraparound, to establish the next working line.

Step 13 Measure off another dimension B (4 13/16 inches) from this working line along ordinates 0 and 0.

Step 14 Draw a line around the pipe along dimension B, using a wraparound, to establish the final working line. *Figure 25* shows the pipe segment working lines.

Step 15 Lay out the ordinate cutbacks on either side of each of the working lines on ordinate lines 4.

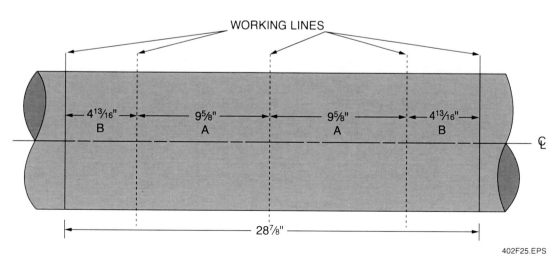

Figure 25 ♦ Pipe segment working lines.

NOTE

Remember that each miter slants in the opposite direction from the previous one. For 12-inch pipe being mitered 15-degrees and divided into 16 ordinates, ordinate number 1 will be ⅝ inches from the working line; ordinate number 2 will be 1³⁄₁₆ inches from the working line; ordinate number 3 will be 1⁹⁄₁₆ inches from the working line; and ordinate number 4 will be 1¹¹⁄₁₆ inches from the working line.

3.6.0 Laying Out Miters Using a Horseshoe

The horseshoe is a shop-built tool that offers an alternative method for laying out miters. Typically, the horseshoe is made from sheet metal, plate steel, or rigid press board. The material that the horseshoe is constructed from is not important as long as it is rigid and will maintain its shape. The inside diameter of the horseshoe must match the outside diameter of the pipe you are working with. The purpose of the horseshoe is to provide a straight edge between the center lines of the pipe at the working line and the center lines of the pipe at the top and bottom of the pipe at the full cutback distance for a given miter. *Figure 27* shows the use of a horseshoe to lay out miters.

As stated earlier, to build a miter, the first consideration must be the radius. Hold the miter as near the same radius as a long radius 90-degree elbow as possible. For this exercise, build a four-piece, 90-degree turn using 4-inch pipe with a radius of 18 inches.

After the radius is determined, the next step is to figure the degree of cuts and the length of each segment. The length of each segment is determined by dividing 90 degrees by the number of welds. In a four-piece 90-degree turn, there are three welds. 90 degrees divided by 3 equals 30 degrees. Since two pipes must be cut to make each 30-degree turn, divide 30 degrees by 2 to obtain the degree of cut for each segment. This gives you 15-degree cuts on each segment. The formula for

Step 16 Use the formulas in the pipefitters' handbooks or calculate the dimensions to mark the cutbacks on the rest of the ordinate lines.

Step 17 Connect the points carefully using a wraparound. *Figure 26* shows the layout of a four-piece, 90-degree, mitered turn.

Step 18 Center-punch marks along either side of ordinates 2 and 2 on each segment.

Step 19 Cut out the segments using an oxyacetylene torch.

Step 20 Bevel the segments, and dress the edges, using a sander or grinder.

Step 21 Fit and align the segments, and have a qualified welder tack-weld them.

Step 22 Check the radius of the completed turn.

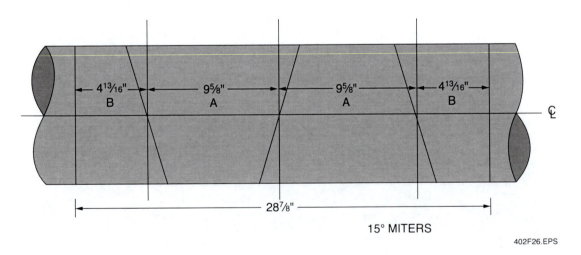

Figure 26 ◆ Layout of four-piece, 90-degree, mitered turn.

figuring the length of each segment when A represents the middle segments and B represents the end segments is:

A = 2 × tan of degree of cut × radius
A = 2 × tan(15) × 18
A = 2 × 0.2679 × 18
A = 2 × 4.822
A = 9.644 or 9⅝ inches
B = A/2
B = 9.644 ÷ 2
B = 4.822 or 4¹³⁄₁₆ inches

After determining the length, you need to figure the cutback of each segment. The formula to do this is:

Cutback = tan of degree of cut × ½ OD of pipe
Cutback = tan(15) × ½ (4.50)
Cutback = 0.2679 × 2.25
Cutback = 0.6027
Cutback = ⅝ inch

Follow these steps to lay out the miters using a horseshoe:

Step 1 Mark the four center lines on the full length of the area of the pipe being used for the miter. Marking the four center lines is the same as dividing the pipe into four ordinates.

Step 2 Lay out the lengths of the two B segments and the two A segments on the pipe.

> **NOTE**
> Refer to *Figure 25*. The A segments should be 9⅝ inches and the B segments should be 4¹³⁄₁₆ inches.

Step 3 Using your wraparound and soapstone, draw a working line around the pipe at the center of the desired turn. This will give you a reference point for measuring and marking the cutback points.

Step 4 Measure to the left of the working line the cutback distance along the center line at the bottom of the pipe and mark the pipe.

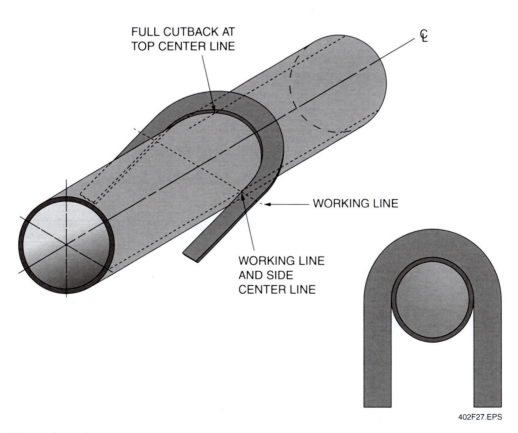

Figure 27 ◆ Using a horseshoe.

Step 5 Measure to the right of the working line the cutback distance along the center line at the top of the pipe and mark the pipe.

Step 6 Place the horseshoe over the top of the pipe so the horseshoe aligns the top center line cutback mark with the intersection of the side center lines and the working line.

Step 7 Mark the pipe along this line using a soapstone marker.

Step 8 Place the horseshoe over the bottom of the pipe so the horseshoe aligns the bottom center line cutback mark with the intersection of the side center lines and the working line.

Step 9 Mark the pipe along this line using a soapstone marker.

Having laid out the miter using the horseshoe, cut the miter using a cutting torch. When cutting the miter, the torch should be aimed at the line on the opposite side of the pipe. After grinding the cut square, back bevel the cuts and grind each cut smooth before rolling the segments 180 degrees from their original position to form the turn. After the center lines are aligned, the pipe must be tacked top and bottom, after checking the pipe for the proper angle with a pipe square.

3.7.0 Mitering a Wye

The procedure for laying out a cutback for a wye is very similar to that for laying out a miter. The working line is drawn on the **branches** and on the header first. The first cutback for the branches is the same as the cutback for the header. The header has two cuts that mirror each other. The branches have a cut that makes the miter with the header and a cut that makes the miter to the other branch (*Figure 28*).

To start out, you need to decide where the intersection of the two pipes will be. The best way to do that is to consider that intersection to be at the center of the header, where the miters meet. Set a working line there, and mark the 4, 8, or 16 axial center lines of the header and the branches at 90 degrees around the pipes just as you would for miter turns. The lines will be where you measure the ordinates for cutbacks.

The first calculation is the miter angle, which will be ½ the intended deviation from the line of the header. The calculation here is the same as that for a miter, except that there is only one weld, so the equation is: degree of bend divided by 2.

If the wye is to be 60 degrees, the degree of deviation from the axis of the header on each side is 30

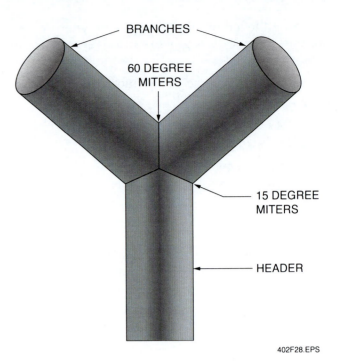

Figure 28 ◆ The wye.

degrees, so the miters between the branches and the header are cut at 15 degrees. The first ordinate, the long cut, on a 4-inch header and branch, is:

Cutback = pipe OD × tangent of miter/2

Cutback = (4.5 × 0.2679)/2

Cutback = 0.6029, or about ⅝ inch

The cutback for the branches of the wye is essentially the same, since the calculation for a 15-degree cutback using the OD of the pipe (12.75 inches) = 1.7075, which also rounds off to 1⅝ inches. Other ordinates are calculated using sine of angle from the working line, just as in calculating an 8- or a 16-ordinate miter. Only half of each piece is mitered in one direction, though on the header, the miters meet in the middle and mirror each other.

So the cutback for the header of a 4-inch, 60-degree wye is ⅝ inch. You aren't trying to figure the radius of a turn, so you don't have to figure the lengths of the miter pieces. The header is the easy piece, just two cuts that mirror each other. The branches have an opposite cut, because one side is matched to the header, while the other side matches up to the other branch. The second cut on the branch is at 60 degrees from the other branch, which turns 30 degrees from the line of the header, so the calculation for the cutback is 60 degrees. That makes your second cut:

Pipe OD × tan 60/2 = (4.5 × 1.73205)/2

Cutback = 3.8971 = 3⅞" when rounded to the nearest sixteenth

To do this one, you would do the layout from the working line for both sides. You have your longest ordinate now, on each side. If this is an 8- or a 16-ordinate layout, the rest of the ordinates are calculated just as they would be for a miter, except that the miters don't form a single straight line.

Lay out the header. Then lay out the short cut on each branch. Then lay out the long cut on each branch. All of the layouts have to be done before doing any cutting, so that any mistakes in the cutting won't spoil the layout for the other cuts.

4.0.0 ◆ FISHMOUTH

The information that you need for a fishmouth, that is, a 90-degree intersection of two pipes, is the same you will need for laterals. You have the ID of the branch, and the OD of the header. One way of visualizing the relationship of the two pieces is with two circles, one the size of the branch ID, and one the size of the header OD. You are transferring the branch ID to the header OD, while making the calculations.

You know the radius of the branch, so you draw the line to the circle of the header. The point where the branch will contact the OD of the header forms a right triangle, one side being the radius of the branch ID and the other (the hypotenuse) being the radius of the header OD *(Figure 29)*. Using the Pythagorean theorem, the sum of the squares of two sides of a right triangle is equal to the square of the hypotenuse. If you subtract the radius squared of the ID of the branch from the radius squared of the OD of the header, you will have the square of the length from the center of the header to the line where the branch will intersect with the header. Take the square root of that number, and you will have a number to use to calculate the longest ordinate of the branch. Subtract this number from the radius of the OD of the header, and the result will tell you how long the long side of the branch will come from the working center line.

If you only needed 4 ordinates, you would have done all the calculations you needed. If you want 8, or 16, you have another set of calculations to perform, for the lengths of the other ordinates. You have one side of a right triangle, and you have the angle between two sides. The side you have is the radius, and that is the hypotenuse of the triangle formed by three sides. The right angle of the triangle is formed by a line drawn to the end of the radius from another radius at 45 degrees to the hypotenuse *(Figure 30)*. That line will equal the radius (the hypotenuse) times the sine of 45 degrees. Now the same set of calculations you used for the long ordinate can be done for the other ordinates, using the side you calculated and the Pythagorean theorem.

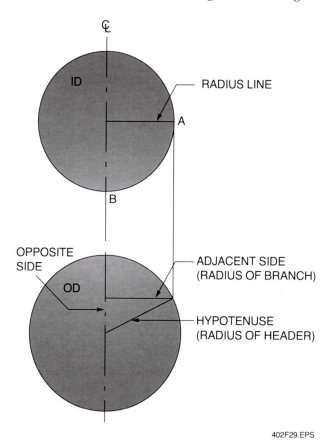

Figure 29 ◆ Calculating the first line.

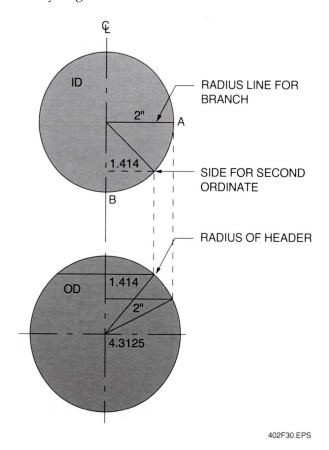

Figure 30 ◆ Calculating the second line.

Let's do an example to get the idea in a real-world form. You have to put together a 4" nominal branch with an 8" standard weight nominal header. Your first line is a 2" radius on the ID of the branch. The OD of the 8" header is 8.625", and the radius would be half of that, or 4.3125". So your first calculation will be side = $\sqrt{(4.3125^2 - 2^2)}$ = 3.8206879... The side is to be subtracted from the radius of the OD of the header, to give you the long ordinate, thus you have ordinate = 4.3125 − 3.8206879... = 0.4918. Rounding to nearest 16th gives you ½". You would draw a working line around the circumference of the branch ½" from the end of the branch, and that would be the end of the cut on the pipe.

Now you want more ordinates so that you can get a smooth, curved cut on the branch. You will figure one more point on the 4" branch, and the line should be smoother. You want a 45-degree angle, so you go to branch radius times sine 45. Now you have line = 2 × sine 45 = 1.4142.

When you plug that into the Pythagorean theorem, you get side = $\sqrt{(4.3125^2 - 1.4142^2)}$ = 4.074... When you subtract that from the radius of the header, you get 4.3125 − 4.074... = 0.23847..., which rounds out to ¼". Use that as the middle point of your layout on the branch and the header, and the two will fit well.

Now let's look at the fit when you are putting a 12" branch on a 16" header. For this, you'll need some more ordinates. The first calculation series is still square root of (header OD radius² − branch ID radius²). That gives you $\sqrt{(8^2 - 6^2)}$ = 5.2915...

Subtract that answer from the radius of the header, and you get 2.7084..., or roughly 2 11/16".

Next, you can get the ordinate lines from the shortest to the longest. You know the zero points on the working line, so start with ordinate 1 = 6 × sin 22.5 degrees = 2.2961... Therefore, $\sqrt{(8^2 - 2.2961^2)}$ = 7.6634..., and subtract that from 8" to get approximately 5/16".

Now you go to 45 degrees, so start with branch radius times sine 45 (6 × sin 45) = 4.24264... Put this answer in the Pythagorean theorem, and you have: $\sqrt{(8^2 - 4.24624...^2)}$ = 6.78233..., or 6 13/16". Subtract that from 8" to get approximately 1.21767 = 1 3/16".

Finally, you have the last ordinate at 6 × sin 67.5 = 5.5432... When you use that in the Pythagorean theorem, you get $\sqrt{(8^2 - 5.432...^2)}$ = 5.7682, or 5¾". Subtract that from 8" to get approximately 2.2318 = 2¼." *Figure 31* shows the two circles, with the triangles shown.

So now we can lay out the set of marks, and cope out the end of the riser, confident that it will fit against the surface of the header.

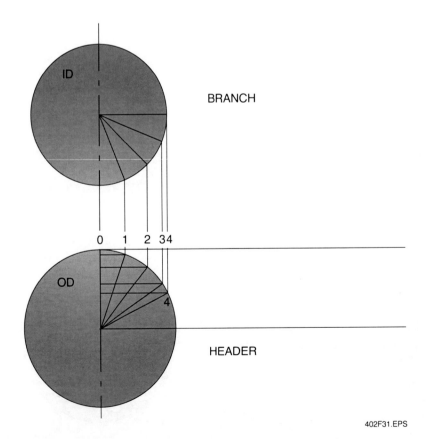

Figure 31 ◆ Finding the rest of the ordinates.

Step 1   Determine where the center of the riser will be on the header.

Step 2   Mark a center line on the top of the header, and extend it on the top of the header past where the outside of the riser will be.

Step 3   Mark a center line on both sides of the riser, using a center head to transfer the line from one side to the other.

Step 4   Measure to the first ordinate you have calculated. This will give you a highest point of the curve you are cutting. Put a working line around the riser at this point.

Step 5   Now you can put the other calculated segment heights and **chord** lengths on the riser. This is done on all four quadrants, and the curves can be drawn with a spline or a wrap. The cuts should be made so that the curve is as smooth as possible, and beveled so that the original cut line is where the edge of the hole is.

Step 6   Put a torpedo level on the side of the riser, and use a framing square on the center line of the header to keep the riser at right angles to the header along that axis.

Step 7   Mark the point where the inside of the bevel touches the header with a soapstone.

Step 8   Cut the hole in the header, and bevel it.

Step 9   Complete the fit-up, and tack-weld it together.

4.1.0 Determining Lateral Dimensions

A lateral is the intersection of two pipes at less than 90 degrees. One-piece laterals are available in a variety of sizes from manufacturers, but pipefitters must often fabricate laterals in the field. The following sections explain how to lay out 45-degree laterals.

When fabricating angled laterals, ordinate reference charts are available. However, these charts may or may not always be available for a given angle, and it is best to be able to calculate the figures for yourself. The equations used for the calculations for risers on 90-degree intersections, supplemented with a little trigonometry, can provide you with the measurements you need for any angle of laterals.

It is necessary to perform the calculations from OD to ID for the laterals, as the insides of the pipes still have to match smoothly, and the OD will match relatively smoothly if the IDs are flush. If the lateral is the same ID on both pieces, the calculation is relatively simple. For a 45-degree lateral of 4-inch Schedule 40 pipe, the two cuts are at a 90-degree angle, meeting at the center of the pipe. One cut is at 22.5 degrees from the axis of the pipe, and the other cut is 67.5 degrees from the axis of the pipe. The fact that the pipe is the same ID lets us know that the pipe will meet in the middle of both pipes, so half of the diameter is the right angle measurement to the intersection of the two cuts. That allows us to calculate the length from the point of intersection in the middle to the point of intersection at the surface for both cuts.

The shorter cut will be calculated by the formula for the tangent of 22.5 degrees, because we know the length to the surface, half the OD. From the point of intersection to the surface is 2.25 inches for a 4-inch Schedule 40 pipe. Tan 22.5 degrees × 2.25 = 0.9319…, or $^{15}/_{16}$ inches. The second, longer cut will be tan 67.5 degrees × 2.25 = 5.43198…, or $5^{7}/_{16}$ inches. These are quick calculations with a scientific calculator.

For a reducing lateral, the calculation is a little more complex, but can still be done with a scientific calculator. For the smaller laterals, such as a 4-inch by 8-inch 45-degree lateral, you could calculate two points on the branch, cut the branch, hold it on the header, and lay out the curve on the header. However, it is more accurate to cut the saddle curve on the branch, so that it fits neatly on the header, and then draw the cut on the header. There are a series of steps for each point. Follow these steps as you refer to *Figure 32*.

Series 1 (Ordinate #1):

Starting with Ordinate #1, the steps are as follows:

Step 1   Line A1 = the radius of the branch ID times the sine of 22.5 degrees = 2.013 × sin 22.5 = 0.7703…

Line B1 is the same size as Line A1.

Step 2   To calculate the length of line C1, use the Pythagorean Theorem, using the outside radius of the header as the length of the hypotenuse, so $C1^2 = (R^2 - B1^2)$. Therefore $C1 = \sqrt{(4.3125^2 - 0.7703^2)} = 4.2431…$

Step 3   Line D1, the height from line B1 to the top of the circle at the longest point is the radius of the header minus line C1: D1 = R − C1 = 0.06936…

Step 4   This line, D1, is the same length as line D1 on the branch ordinates. Divide D1 by the sine of the angle of the lateral to get the length of line E1: .06936…/sin 45 = 0.09809… (Note E1 and E7 are the same length.)

Step 5 Find line F1 by multiplying the radius of the branch ID by sin 67.5, F1 = sin 67.5 × 2.013 = 1.85976… (Note: F1 and F7 are the same length.)

Step 6 Subtract the inside radius of the branch from F1 to get G1: 2.013 − 1.85976… = 0.1532…

Step 7 Line H1 = line G1 divided by the tangent of the angle of the lateral = 0.1532…/tan 45 = 0.1532…, because the tangent of 45 degrees is 1.

Step 8 To find I1 add line H1 to line E1: 0.09814… + 0.1532… = 0.25137… or ¼".

Now figure Ordinate #7. All the steps are the same until you get to Step 6.

Step 6 Add the inside radius of the branch to F7 to get G7: 2.013 + 1.85976…= 3.8727…

Step 7 Line H7 = line G7 divided by the tangent of the angle of the lateral = 3.8727…/ tan 45 = 3.8727…because the tangent of 45 degrees is 1.

Step 8 To find I7 add Line H7 to line E7: 0.09814… + 3.8727… = 3.9709…. or 4".

Series 2 (Ordinate #2):

Step 1 Line A2 = radius of branch times sin 45 degrees = 2.013 × sin 45 = 1.4234…

Line B2 = Line A2

Step 2 Line $C2^2$ = radius of $header^2$ − line $B2^2$. Therefore, C2 = $\sqrt{(4.3125^2 - 1.4234^2)}$ = 4.0708…

Step 3 Line D2 = radius of header − line C2 = 4.3125 − 4.0708… = 0.2417…

Step 4 Line E2 = D2 divided by the sine of the angle of the lateral = 0.2417… / sin 45 degrees = 0.3418… (Note E2 and E6 are the same length.)

Step 5 Line F2 = radius of branch × sin 45 = 2.013 × sin 45 = 1.423… (Note F2 and F6 are the same length.)

Step 6 Line G2 = radius of branch − line F2 = 2.013 − 1.423… = 0.5895…

Step 7 Line H2 = line G2/tan 45 degrees = 0.5895…

Step 8 Line I2 = Line H2 + line E2 = 0.5895… + 0.3418… = 0.9313… or ¹⁵⁄₁₆".

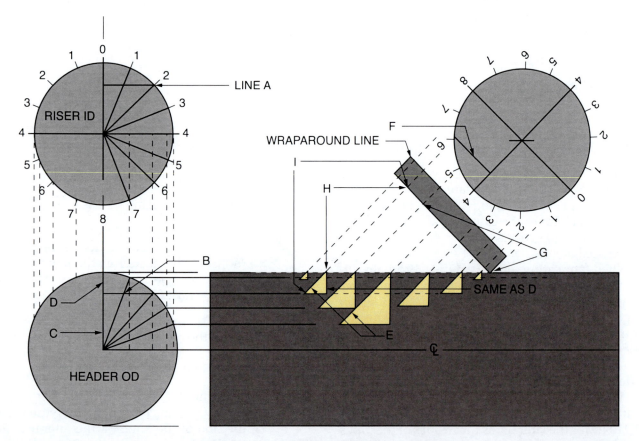

FINDING THE 16 ORDINATES FOR A LATERAL

Figure 32 ◆ Layout for a lateral.

Now figure Ordinate #6. All the steps are the same until you get to Step 6:

Step 6 Add the inside radius of the branch to F6 to get G6: $2.013 + 0.3418... = 2.3548...$

Step 7 Line H6 = line G6 divided by the tangent of the angle of the lateral = $2.3548.../\tan 45 = 2.3548...$

Step 8 To find I7 add line H7 to line E7: $0.09814... + 2.3548... = 2.4529...$ or $3\frac{3}{4}"$.

Series 3 (Ordinate #3):

Step 1 Line A3 = radius of branch times sin 67.5 = $2.013 \times \sin 67.5 = 1.8597...$

Line B3 = Line A3.

Step 2 Line $C3^2$ = radius of header^2 − Line $B3^2$ = $4.312^2 - 1.8597...^2 = 15.139...$

Line C3 = $\sqrt{15.139...} = 3.8909...$

Step 3 Line D3 = radius of header − line C3 = $4.3125 - 3.8909 = 0.42162...$

Step 4 Line E3 = D3 divided by the sine of the angle of the lateral = $0.42162.../\sin 45 = 0.5962...$ (Note E3 and E5 are the same length.)

Step 5 Line F3 = radius of branch × sin 22.5 = $2.013 \times \sin 22.5$ degrees = $0.7703...$ (Note F3 and F5 are the same length.)

Step 6 Line G3 = radius of branch − line F3 = $2.013 - 0.7703... = 1.2426...$

Step 7 Line H3 = Line G3 / tan 45 degrees = $1.2426...$

Step 8 Line I3 = line H3 + line E3 = $1.2426... + 0.5962... = 1.8389...$ = $1\frac{13}{16}"$.

Now figure Ordinate #5. All the steps are the same until you get to Step 6:

Step 6 Add the inside radius of the branch to F5 to get G5: $2.013 + 0.7703... = 2.7833...$

Step 7 Line H5 = line G5 divided by the tangent of the angle of the lateral = $2.7833.../\tan 45 = 2.7833...$

Step 8 To find I5 add Line H5 to line E5: $0.5962... + 2.7833... = 3.3795...$ or $3\frac{3}{8}"$.

Series 4 (Ordinate #4):

Step 1 Line A4 = radius of branch times sin 90 degrees = $2.013 \times \sin 90 = 2.013$

Line B4 = line A4

Step 2 Line $C4^2$ = radius of header^2 − line $B4^2$ = $4.3125^2 - 2.013^2 = 14.5454...$

C4 = $\sqrt{14.5454...} = 3.8138...$

Step 3 Line D4 = radius of header − line C4 = $4.3125 - 3.8138... = .4986...$

Step 4 Line E4 = D4 divided by the sine of the angle of the lateral = $.4986.../\sin 45 = .7051...$

Step 5 Line F4 = radius of the branch × sine 90 degrees = $2.013 \times \sin 90$ degrees = 2.013

Step 6 Line G4 = radius of branch = 2.013 (Note that we have not added anything to the branch radius. Point 4 is on the radius; we won't add to it here.)

Step 7 Line H4 = Line G4 / tan 45 degrees = 2.013

Step 8 Line I4 = line H4 + line E4 = $2.013 + .7051... = 2.7181...$ or $2\frac{11}{16}"$

Series 5 (Ordinate #8):

Step 1 Line A8 is zero.

Since A8 is equal to B8, it is also zero.

Since B8 is zero, C8 is equal to the radius of the header OD and there is no D8.

Since there is no D8, there is no E8 as well, since it equals zero.

Step 2 Line F8 is equal to the radius of the branch; therefore, F8 equals 2.013".

Step 3 Line G8 is equal to F8 + the radius of the branch. G8 = $2.013 + 2.013 = 4.026"$

Step 4 Line H8 = G8 divided by the tangent of the angle of the lateral. H8 = $4.026/\tan 45$ degrees = $4.026"$.

Step 5 Line I8 = H8 + E8 = $4.026 + 0 = 4.026$ or 4".

When references are available, they make the job of laying out laterals easier and quicker than laying them out on a flat surface and then transferring them to the pipe. For this exercise, you will lay out a 45-degree lateral, using 4-inch carbon steel pipe. Follow these steps to lay out a 45-degree lateral, using references:

Step 1 Obtain a 45-degree lateral reference chart, such as the one shown in *Figure 33*.

Step 2 Measure from the end of the pipe 5 or 6 inches, and mark a straight line around the pipe using a wraparound. This line is the base line.

45° Laterals Standard Weight Pipe Risers Eight Ordinates													
SIZE OF RISER	SIZE OF HEADER												ORDINATE NO.
	3"	4"	6"	8"	10"	12"	14"	16"	18"	20"	22"	24"	
3"	1	¹³⁄₁₆	¹¹⁄₁₆	⅝	⅝	⁹⁄₁₆	⁹⁄₁₆	⁹⁄₁₆	⁹⁄₁₆	⁹⁄₁₆	½	½	1
	2¹³⁄₁₆	2⅜	2¹⁄₁₆	1¹⁵⁄₁₆	1⅞	1¹³⁄₁₆	1¾	1¾	1¾	1¹¹⁄₁₆	1¹¹⁄₁₆	1¹¹⁄₁₆	2
	3⅛	3	2⅞	2¹³⁄₁₆	2¾	2¾	2¾	2¾	2¹¹⁄₁₆	2¹¹⁄₁₆	2¹¹⁄₁₆	2¹¹⁄₁₆	3
	3¹⁄₁₆	3¹⁄₁₆	3	3¹⁄₁₆	3¹⁄₁₆	3¹⁄₁₆	3¹⁄₁₆	3¹⁄₁₆	3¹⁄₁₆	3¹⁄₁₆	3¹⁄₁₆	3¹⁄₁₆	4
4"	—	1⁵⁄₁₆	1¹⁄₁₆	¹⁵⁄₁₆	⅞	¹³⁄₁₆	¹³⁄₁₆	¾	¾	¾	¾	¹¹⁄₁₆	1
	—	3¾	3	2¹¹⁄₁₆	2⁹⁄₁₀	2½	2⁷⁄₁₆	2⅜	2¹⁵⁄₁₆	2¹⁵⁄₁₆	2¼	2¼	2
	—	4⅛	3⅞	3¾	3¹¹⁄₁₆	3¹¹⁄₁₆	3⅝	3⅝	3⅝	3⁹⁄₁₆	3⁹⁄₁₆	3⁹⁄₁₆	3
	—	4	4	4	4	4	4	4	4	4	4	4	4
6"	—	—	2	1¹¹⁄₁₆	1½	1⁷⁄₁₆	1⅜	1⁵⁄₁₆	1¼	1³⁄₁₆	1³⁄₁₆	1³⁄₁₆	1
	—	—	5¹³⁄₁₆	4¹³⁄₁₆	4⅜	4⅛	4	3⅞	3¾	3¹¹⁄₁₆	3⅝	3⁹⁄₁₆	2
	—	—	6⁵⁄₁₆	6	5⁵⁄₁₆	5¹¹⁄₁₆	5⅝	5⁹⁄₁₆	5⁹⁄₁₆	5½	5½	5⁵⁄₁₆	3
	—	—	6¹⁄₁₆	6¹⁄₁₆	6¹⁄₁₆	6¹⁄₁₆	6¹⁄₁₆	6¹⁄₁₆	6¹⁄₁₆	6¹⁄₁₆	6¹⁄₁₆	6¹⁄₁₆	4
8"	—	—	—	2⅝	2⁵⁄₁₆	2⅛	2	1⅞	1¹³⁄₁₆	1¾	1¹¹⁄₁₆	1⅝	1
	—	—	—	7¾	6½	6	5¾	5½	5⁵⁄₁₆	5³⁄₁₆	5¹⁄₁₆	4¹⁵⁄₁₆	2
	—	—	—	8⁵⁄₁₆	7¹⁵⁄₁₆	7¾	7⅝	7⁹⁄₁₆	7⁷⁄₁₆	7⅜	7⁵⁄₁₆	7⁵⁄₁₆	3
	—	—	—	8	8	8	8	8	8	8	8	8	4

Figure 33 ◆ 45-degree lateral reference chart.

Step 3 Divide the outside circumference of the pipe by eight to obtain eight ordinates.

Step 4 Mark the eight ordinates on the pipe, 0 to 4 on each side, as shown in *Figure 34*.

Step 5 Obtain the ordinate 1 cutback measurement from *Figure 33*.

Step 6 Measure 1⁵⁄₁₆ inch from the base line along ordinate 1 and mark the pipe.

Step 7 Obtain the ordinate 2 cutback measurement from *Figure 33*.

Step 8 Measure 3¾ inches from the base line along ordinate 2 and mark the pipe.

Step 9 Obtain the ordinate 3 cutback measurement from *Figure 33*.

Step 10 Measure 4⅛ inches from the base line along ordinate 3 and mark the pipe.

Step 11 Obtain ordinate 4 cutback measurement from *Figure 33*.

Step 12 Measure 4 inches from the base line along ordinate 4 and mark the pipe.

Step 13 Place a wraparound on the lateral and mark a line connecting ordinates 1, 0, and 1, using a soapstone.

Step 14 Place a wraparound on the lateral and mark a line connecting ordinates 1, 2, and 3, using a soapstone.

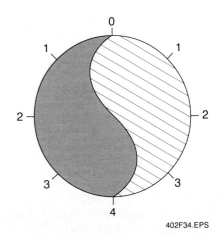

Figure 34 ◆ Numbering ordinates.

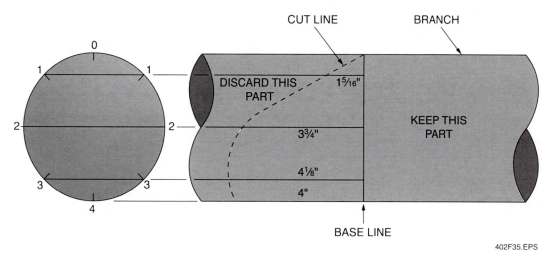

Figure 35 ♦ Lateral ready to cut.

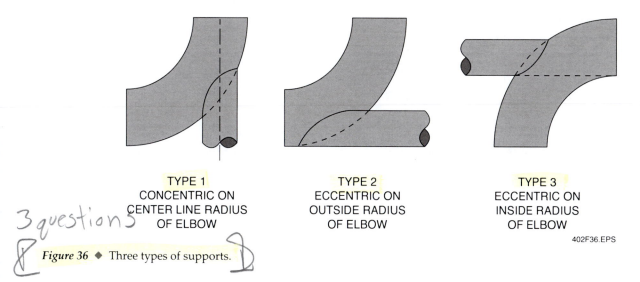

Figure 36 ♦ Three types of supports.

Step 15 Place a wraparound on the lateral and mark a line connecting ordinates 3, 4, and 3, using a soapstone. *Figure 35* shows the lateral ready to cut.

Step 16 Cut the lateral along the cut line, using a cutting torch, keeping the torch at a 90-degree angle to the pipe.

Step 17 Clean the slag off the lateral using a grinder.

Step 18 Place the lateral on the header and use a torpedo level to establish a 45-degree angle.

Step 19 Mark around the outside diameter of the lateral on the header using a soapstone.

Step 20 Cut along the inside of the mark on the header using a cutting torch. Be sure to make the cut at a 90-degree angle with the header.

Step 21 Grind off any slag to clean the header.

Step 22 Place the lateral on the header and check the fit.

Step 23 Make any necessary adjustments and bevel the lateral and header.

Step 24 Fit up the lateral to the header using spacer wires to obtain the correct welding gap; verify that the header is level and use a torpedo level to establish a 45-degree angle.

Step 25 Tack-weld the lateral to the header.

Step 26 Verify that the lateral is still at a 45-degree angle to the header.

Step 27 Complete the weld.

5.0.0 ♦ FABRICATING DUMMY LEGS AND TRUNIONS OUT OF PIPE

Pipefitters are sometimes required to fabricate project-made supports known as dummy legs or trunions to help hold a piping system in place. Dummy legs and trunions are fabricated out of pipe and are usually welded directly to the pipe on a 90-degree, long-radius elbow and are then welded to an anchor plate on the floor, wall, or ceiling. Dummy legs provide horizontal support, and trunions provide vertical support. There are three types of dummy legs and trunions on 90-degree, long-radius elbows. The type of dummy leg or trunion used depends on the area of the elbow to which the support is attached. *Figure 36* shows three types of supports.

The layout for each type of support is the same although each type requires different dimensions. The dimensions for each type of support can be found in *The Pipe Fitters Blue Book*. The procedure for laying out the supports is very similar to the procedure for laying out 45-degree laterals, using reference charts. *Figure 37* shows the layout for a support on the back of an elbow.

For this procedure, you will fabricate a type 1 support, using 3-inch, carbon steel, 40-weight pipe that will support a 6-inch, long-radius, 90-degree elbow. The center line of the support is in line with the center line of the elbow. The support must be cut so that the center line of the pipe is 2 feet, 6 inches above grade elevation. A 6-inch base and ½-inch baseplate will be used with the support. The 90-degree elbow is part of a spool consisting of the elbow and a straight length of pipe. *Figure 38* shows the support.

Follow these steps to fabricate the support:

Step 1   Obtain the support detail sheet and specifications.

Step 2   Review the support detail sheet to determine the size of pipe to be used to fabricate the support.

Step 3   Determine what type of support the specifications require.

Step 4   Obtain the pipe to be used for the support.

Step 5   Secure the spool with the elbow in a pipe vise and jack stands.

Step 6   Obtain a reference chart for the type and size support needed. *Figure 39* shows a reference chart for a type 1 concentric support on the back of a 90-degree, long-radius elbow.

Step 7   Locate the ordinate measurements for the pipe being used on the reference chart and find the longest measurement.

Step 8   Measure 12 inches from the end of the 3-inch pipe and mark the pipe at this point. The dimension of 12 inches is used to give you a point past the longest ordinate measurement.

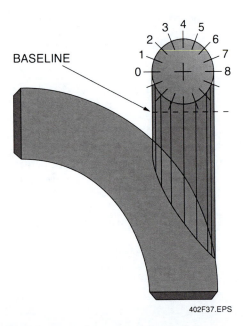

Figure 37 ♦ Layout of support on back of elbow.

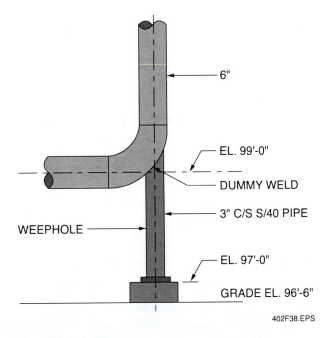

Figure 38 ♦ Support.

Step 9 Place a wraparound on the pipe at this mark and scribe a line around the pipe. This line is the base line.

Step 10 Divide the outside circumference of the pipe by 16 to obtain 16 ordinates.

Step 11 Mark the 16 ordinate lines on the pipe 0 to 8 on each side using an angle iron as a straight edge.

Step 12 Obtain the ordinate 1 cutback measurement from the reference chart.

Step 13 Measure 8⅜ inches from the base line along ordinate 1 and mark the pipe.

Step 14 Obtain the ordinate 2 cutback measurement from the reference chart.

Step 15 Measure 8³⁄₁₆ inches from the base line along ordinate 2 and mark the pipe.

Step 16 Continue to measure and mark the ordinates on the pipe according to the reference chart.

	Size Of Elbow							
	2"	3"	4"	6"	8"	10"	12"	NO.
2" PIPE	2⁵⁄₁₆	3¹³⁄₁₆	5¹⁄₁₆	8⅝	11¹⁵⁄₁₆	15³⁄₁₆	18½	0
	2⁷⁄₁₆	3¹⁵⁄₁₆	5½	8¹¹⁄₁₆	12	15¼	18⁹⁄₁₆	1
	2¹³⁄₁₆	4³⁄₁₆	5¹³⁄₁₆	9	12³⁄₁₆	15½	18⅞	2
	3⅜	4¹¹⁄₁₆	6¼	9⅜	12¹¹⁄₁₆	15¹⁵⁄₁₆	19¼	3
	4¹⁄₁₆	5³⁄₁₆	6¹¹⁄₁₆	9¹³⁄₁₆	13⅛	16⅜	19¹¹⁄₁₆	4
	4½	5⁹⁄₁₆	7⅛	10¼	13⁹⁄₁₆	16¹³⁄₁₆	20⅞	5
	4¾	5⅞	7⁷⁄₁₆	10⁹⁄₁₆	13⅞	17⅛	20⁷⁄₁₆	6
	4⅞	6¹⁄₁₆	7⅝	10¹³⁄₁₆	14⅛	17⅜	20¹¹⁄₁₆	7
	4⅞	6⅛	7¹¹⁄₁₆	10⅞	14³⁄₁₆	17⁷⁄₁₆	20¾	8
3" PIPE	—	3½	5¹⁄₁₆	8¹³⁄₁₆	11½	14¹¹⁄₁₆	18	0
	—	3¹¹⁄₁₆	5¼	8⅜	11⅝	14⅞	18³⁄₁₆	1
	—	4¼	5¾	8¹³⁄₁₆	12¹⁄₁₆	15¼	18⁹⁄₁₆	2
	—	5⅛	6⁷⁄₁₆	9⁷⁄₁₆	12¹¹⁄₁₆	15⅞	19³⁄₁₆	3
	—	6⅛	7¼	10³⁄₁₆	13⅜	16⁹⁄₁₆	19¹³⁄₁₆	4
	—	6¹³⁄₁₆	7¹⁵⁄₁₆	10¹³⁄₁₆	14	17³⁄₁₆	20½	5
	—	7³⁄₁₆	8¾	11¼	14½	17¹¹⁄₁₆	21	6
	—	7¹⁵⁄₁₆	8⁹⁄₁₆	11⁹⁄₁₆	14¹³⁄₁₆	18	21³⁄₁₆	7
	—	7⅜	8⁵⁄₁₆	11⅝	14⅞	18¹⁄₁₆	21⁷⁄₁₆	8
4" PIPE	—	—	4¾	7⅞	11⅛	14⁵⁄₁₆	17⅜	0
	—	—	5	8¹⁄₁₆	11⁵⁄₁₆	14½	17¹³⁄₁₆	1
	—	—	5¾	8¹¹⁄₁₆	11⅞	15¹⁄₁₆	18⅜	2
	—	—	6¹⁵⁄₁₆	9⅝	12¾	15⅞	19⅛	3
	—	—	8⅜	10⅝	13¹¹⁄₁₆	16¹³⁄₁₆	20¹⁄₁₆	4
	—	—	9⅜	11½	14⁹⁄₁₆	17⅝	20⅞	5
	—	—	9¹³⁄₁₆	12¹⁄₁₆	15³⁄₁₆	18¼	21⁹⁄₁₆	6
	—	—	10	12⅜	15½	18¹¹⁄₁₆	21¹⁵⁄₁₆	7
	—	—	10¹⁄₁₆	12½	15⅝	18¹³⁄₁₆	22¹⁄₁₆	8
6" PIPE	—	—	—	7¼	10⅜	13½	16¾	0
	—	—	—	7⅝	10¾	13¹³⁄₁₆	17⁵⁄₁₆	1
	—	—	—	8¾	11¾	14¾	18	2
	—	—	—	10⅝	13¼	16⅛	19⁵⁄₁₆	3
	—	—	—	12¹⁵⁄₁₆	14⅞	17⅝	20¾	4
	—	—	—	14⁷⁄₁₆	16¼	18¹⁵⁄₁₆	22	5
	—	—	—	15¹⁄₁₆	17¹⁄₁₆	19¹³⁄₁₆	22¹⁵⁄₁₆	6
	—	—	—	15³⁄₁₆	17⁹⁄₁₆	20⅜	23½	7
	—	—	—	15⅜	17¹¹⁄₁₆	20½	23¹¹⁄₁₆	8

Figure 39 ♦ Type 1 support reference chart.

Step 17 Connect all of the ordinate marks using the wraparound.

Step 18 Cut the support along the cut line, using a cutting torch, keeping the torch at a 90-degree angle to the pipe.

Step 19 Clean the support to remove any slag, using a grinder.

NOTE
It is frequently the case that a support leg is required to have a weep hole. Check local standards to be sure, but if it is not forbidden, drill a small hole before welding the support leg to the elbow.

Step 20 Align the support with the elbow on the pipe, using a framing square and levels, as shown in *Figure 40*.

Step 21 Fit up the branch to the header using spacer wires to obtain the correct welding gap.

Step 22 Tack-weld the branch to the header.

Step 23 Verify that the support is square and properly aligned with the center line of the elbow.

Step 24 Have the welder weld the support to the elbow.

Step 25 Transfer a mark from the center line of the pipe to the support.

Step 26 Determine the length that the support needs to be cut. Subtract the height of the base and the thickness of the baseplate from the measurement between the center line of the pipe and the grade elevation to determine the length. For this example, the length between the baseplate and the center line of the pipe is 23½ inches.

Step 27 Measure the support from the center line mark and mark the support at this point.

Step 28 Scribe a line around the support at this mark using a wraparound.

Step 29 Cut the support at this line using a cutting torch.

Step 30 Clean the slag from the cut pipe end.

Step 31 Square the baseplate to the end of the support.

Step 32 Weld the baseplate to the pipe end.

6.0.0 ◆ LAYING OUT LATERALS AND SUPPORTS WITHOUT USING REFERENCES

Laying out laterals and supports can also be performed when references are not available. To accomplish this, you must first lay out the cut on a piece of paper or other flat surface, and then transfer the cut line measurements to the pipe to be cut. This section explains the procedures for laying out 45-degree laterals and dummy leg supports without using reference charts.

6.1.0 Laying Out Laterals

When references are not available, laying out laterals can be accomplished by laying out the lateral on a flat surface and then transferring this to the pipe. In order to accurately perform the following procedure, you will need some drafting tools: a

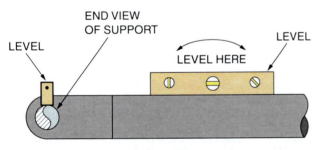
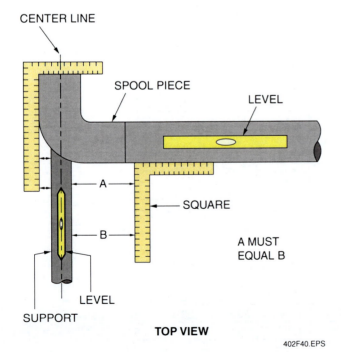

Figure 40 ◆ Aligning support to elbow.

compass, a 45-degree right triangle, dividers, and a ruler. A T-square would also be helpful. Follow these steps to layout a 45-degree lateral without using reference charts:

Step 1 Measure the exact inside and outside diameter of the pipe you are using.

Step 2 Draw a vertical straight line that is at least 2½ times the outside diameter of the pipe on a sheet of drafting paper. This line will become the vertical center line of two circles representing the inside diameter branch and outside diameter header of the pipe.

Step 3 Use a compass to draw a circle that is the exact size of the ID of the pipe, using a point on the straight line as the center.

Step 4 Use a compass to draw a circle that is the exact size of the OD of the pipe, using a point on the straight line below the ID circle as the center (*Figure 41*).

Step 5 Use the right triangle or T-square to draw a 90-degree line from the center of the ID circle to the outside edge of the circle (*Figure 42*).

Step 6 Using dividers, find the center between points A and B on the circumference of the circle and draw a line from the center of the circle to this point on the edge of the circle.

Step 7 Using dividers, find the center between points A and C and B and C and draw a line from the center of the circle to these points on the edge of the circle (*Figure 43*).

Step 8 Transfer points A, B, C, D, and E to the circumference of the pipe OD circle. Make sure that these lines are parallel to the straight line drawn in Step 2.

Step 9 Label the points where these lines intersect the circumference of the OD circle starting with the vertical center line being 0 and numbering 1, 2, 3, and 4 around the circle (*Figure 44*).

Step 10 Draw lines perpendicular to the vertical line at point 0 and at the center point of the pipe OD circle (*Figure 45*).

Step 11 Draw a line 45 degrees off of the two horizontal lines.

Step 12 Draw another ID circle half toward the end of the 45-degree line and mark the center line of this circle half with a line that is perpendicular to the 45-degree line (*Figure 46*).

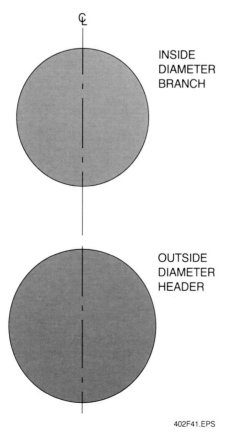

Figure 41 ♦ Step 4.

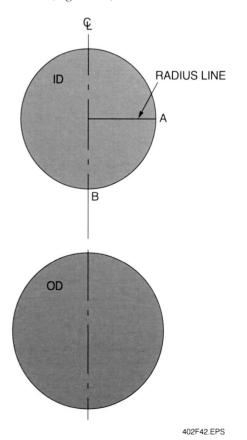

Figure 42 ♦ Step 5.

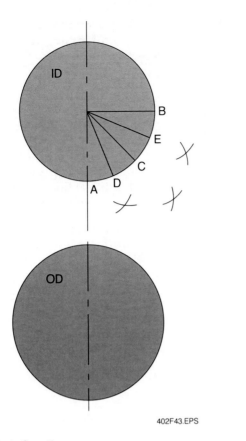

Figure 43 ◆ Step 7.

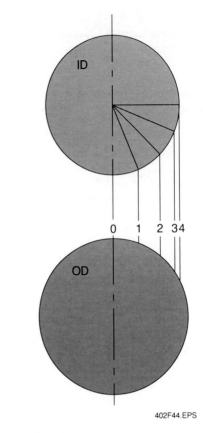

Figure 44 ◆ Step 9.

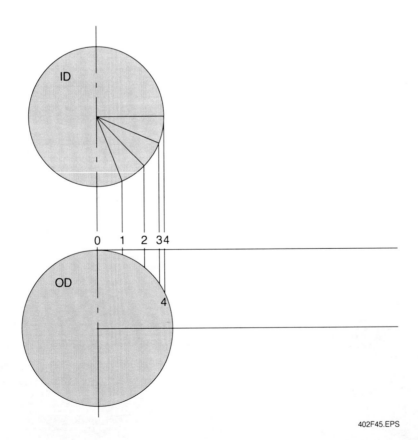

Figure 45 ◆ Step 10.

2.34   PIPEFITTING   ◆   LEVEL FOUR

Step 13 Using the dividers, divide this half circle into four equal parts.

Step 14 Using the dividers, divide each part in half so that the half circle is divided into eight equal parts.

Step 15 Label these sections with ordinate numbers 0 through 8 (*Figure 47*).

Step 16 Draw lines perpendicular to the vertical line off of the OD circle points 1, 2, 3, and 4 (*Figure 48*).

Step 17 Transfer the points from the ID circle half sections to the horizontal lines coming off of the OD circle points 0, 1, 2, 3, and 4 (*Figure 49*).

Step 18 Draw a wraparound line perpendicular to these lines. This gives you a reference point for measuring the cutback distances.

Step 19 Measure from the wraparound line down to the points where the transfer lines meet the ordinate lines (*Figure 50*). These measurements give you the ordinate cutback measurements for laying out your lateral.

Step 20 Lay out sixteen ordinates, labeled 0 through 8, and a working line on the lateral pipe.

Step 21 Use the measurements found above to lay out the cut line on the lateral pipe.

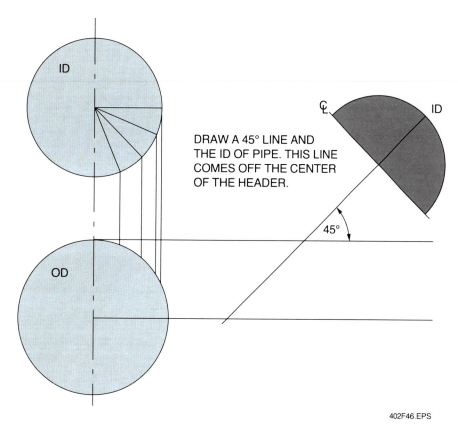

Figure 46 ◆ Step 12.

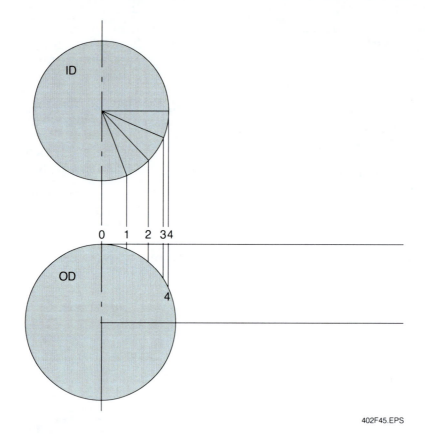

Figure 47 ◆ Step 15.

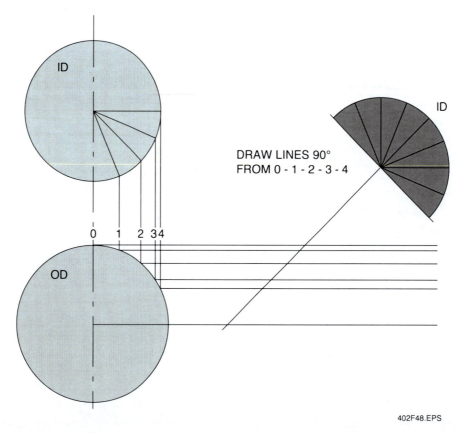

Figure 48 ◆ Step 16.

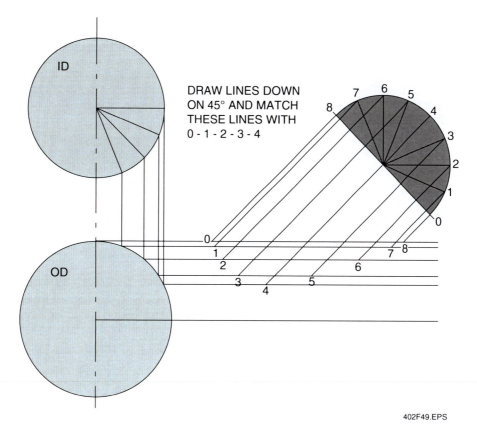

Figure 49 ◆ Step 17.

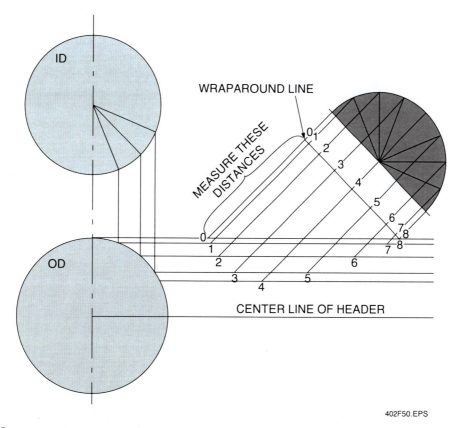

Figure 50 ◆ Step 19.

MODULE 08402-07 ◆ ADVANCED PIPE FABRICATION 2.37

6.2.0 Laying Out Supports

When references are not available, laying out pipe supports, such as dummy legs, can be accomplished by laying out the support on a sheet of paper and then transferring this to the pipe. In order to accurately perform the following procedure, you will need some drafting tools: a compass, a 45-degree right triangle, dividers, and a ruler. A T-square would also be helpful. Follow these steps to lay out an ell support without using reference charts.

Step 1 Measure the exact inside and outside diameter of the ell and the pipe you are using.

Step 2 Draw a side view of the ell to exact dimensions on a sheet of drafting paper (*Figure 51*).

Step 3 Extend the horizontal center line of the ell through the center of the ell to the left edge of the paper.

Step 4 Using the center line of the ell as a center reference, draw a circle representing the outside diameter header of the ell (*Figure 52*).

Step 5 Draw a vertical center line through the midpoint of this circle to the top edge of the paper.

Step 6 Using the center line of the OD circle as a center reference, draw a circle representing the inside diameter of the support (*Figure 53*).

Step 7 Use the right triangle or T-square to draw a 90-degree line from the center of the ID circle to the outside edge of the circle (*Figure 54*).

Step 8 Using dividers, find the center between points A and B on the circumference of the circle and draw a line from the center of the circle to this point on the edge of the circle.

Step 9 Using dividers, find the center between points A and C and B and C and draw a line from the center of the circle to these points on the edge of the circle (*Figure 55*).

Step 10 Transfer points A, B, C, D, and E to the circumference of the pipe outside diameter header circle.

NOTE
Make sure that these lines are parallel to the straight line drawn in Step 7.

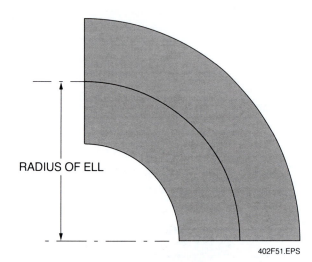

Figure 51 ◆ Step 2.

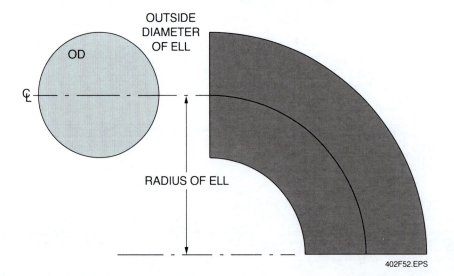

Figure 52 ◆ Step 4.

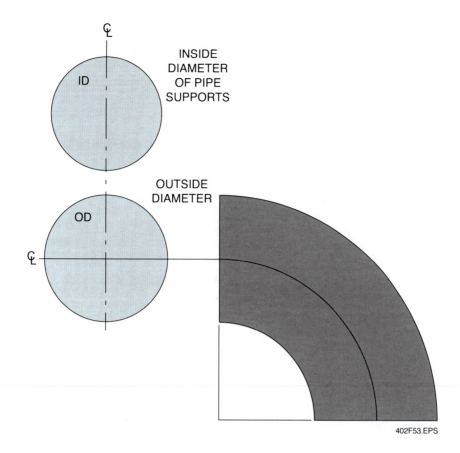

Figure 53 ◆ Step 6.

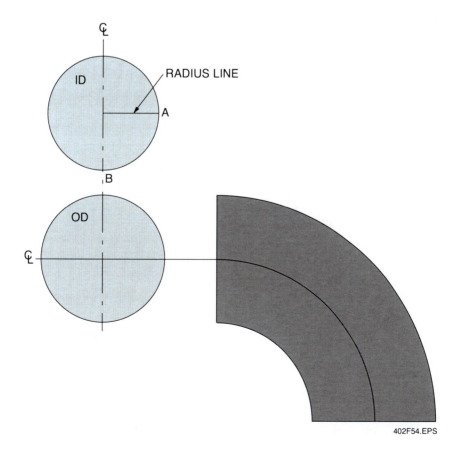

Figure 54 ◆ Step 7.

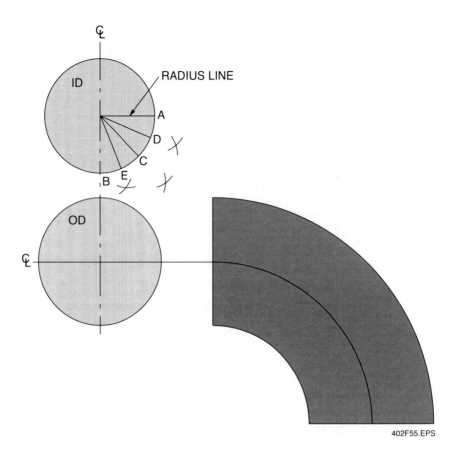

Figure 55 ◆ Step 9.

Step 11 Label the points where these lines intersect the circumference of the outside diameter header circle starting with the vertical center line being 0 and numbering 1, 2, 3, and 4 around the circle (*Figure 56*).

Step 12 Transfer points 0, 1, 2, 3, and 4 to the face of the ell.

Step 13 Using a compass from the center point of the ell radius, transfer lines 0, 1, 2, 3, and 4 the length of the ell (*Figure 57*).

Step 14 Extend the vertical center line of the ell through the center of the ell to the top edge of the paper.

Step 15 Draw another ID circle half toward the end of the vertical ell center line and mark the center line of this circle half with a line that is perpendicular to the vertical line (*Figure 58*).

Step 16 Using the dividers, divide this half circle into four equal parts.

Step 17 Using the dividers, divide each part in half so that the half circle is divided into eight equal parts (*Figure 59*).

Step 18 Label each point on the half circle circumference with ordinates 0 through 8.

Step 19 Transfer points 0 through 8 vertically down to the ell lines drawn in Step 12 (*Figure 60*).

Step 20 Draw a wraparound line perpendicular to these lines. This gives you a reference point for measuring the cutback distances.

Step 21 Measure from the wraparound line down to the points where the transfer lines meet the ordinate lines. These measurements give you the ordinate cutback measurements for laying out the pipe support (*Figure 61*).

Step 22 Lay out 16 ordinates and a working line on the pipe support.

Step 23 Use the measurements found above to lay out the cut line on the pipe support.

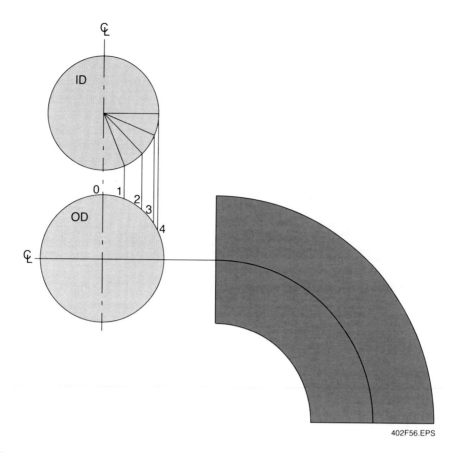

Figure 56 ♦ Step 11.

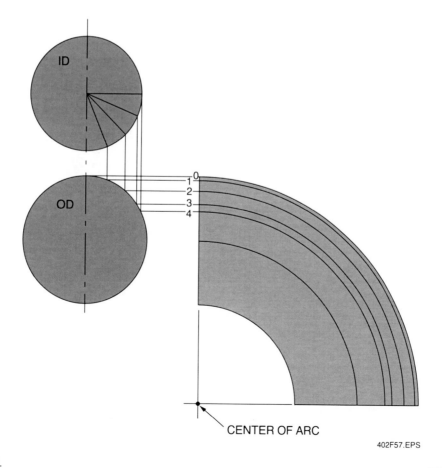

Figure 57 ♦ Step 13.

MODULE 08402-07 ♦ ADVANCED PIPE FABRICATION 2.41

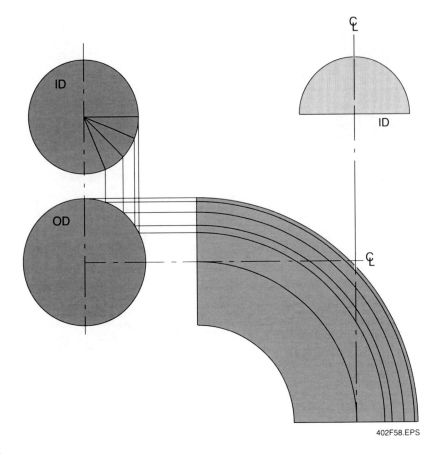

Figure 58 ◆ Step 15.

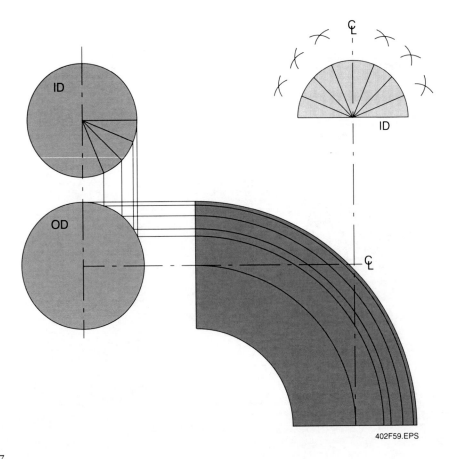

Figure 59 ◆ Step 17.

2.42 PIPEFITTING ◆ LEVEL FOUR

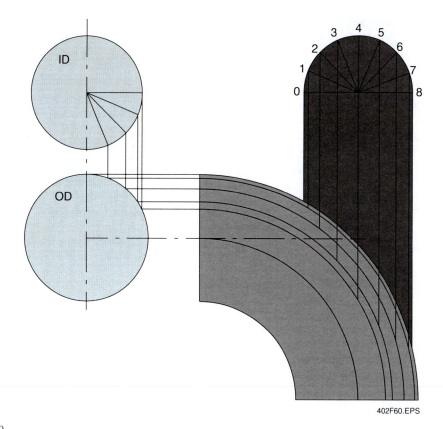

Figure 60 ◆ Step 19.

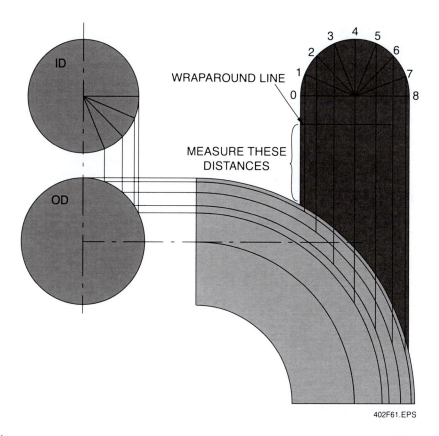

Figure 61 ◆ Step 21.

Review Questions

1. The sum of the angles of a triangle is _____ degrees.
   a. 90
   b. 180
   c. 270
   d. 360

2. In order to solve any right triangle, you must know the length of two sides or the _____.
   a. degrees of one angle
   b. degrees of two angles and the length of one side
   c. length of all three sides
   d. length of one side

3. To find the length of travel for a simple 45-degree offset when the set is known, multiply the set by _____.
   a. 0.7071
   b. 1
   c. 1.414
   d. 2

4. The distance from the center of a vessel to the starting point of an offset is equal to the distance from the center of the vessel to the center line of the pipe times the _____.
   a. sine of the offset angle
   b. tangent of the offset angle
   c. tangent of half the offset angle
   d. cosine of the offset angle

5. The set and travel of all three lines in an equal-spread offset is the same, but the _____ are different.
   a. runs
   b. angles of offset
   c. starting points
   d. tangents of the angles

6. For 45-degree equal-spread offsets, the difference in starting points can be obtained by multiplying the distance between adjacent lines by _____.
   a. 0.4142
   b. 0.7071
   c. 1.414
   d. 2.414

7. For three-line unequal-spread offsets, you need to know the center-to-center distance between the lines before the offset _____.
   a. and not after the offset
   b. and after the offset
   c. except for line 2
   d. except for line 1

8. For a three-line unequal-spread 45-degree offset, lines 1 and 2 are 16 inches apart, and lines 2 and 3 are 20 inches apart. At the other end, lines 1 and 2 are 16 inches apart, and lines 2 and 3 are 32 inches apart. The starting point for line 3 is _____ inches past the start for line 1.
   a. 6.34
   b. 14.91
   c. 20
   d. 31⅞

9. The center-to-center lengths for pipes in a coil can be obtained by multiplying 2 times the radius of the coil by the _____.
   a. cosine of the angle
   b. sine of the angle
   c. tangent of ½ the angle
   d. number of pipes in the coil

10. Pipe laterals that are 4 inches through 10 inches are laid out with _____ ordinates.
    a. 4
    b. 8
    c. 16
    d. 32

11. The outside diameter of the pipe is multiplied by the _____ to obtain the cutback for a miter.
    a. sine of the angle
    b. tangent of the angle, then divide by two
    c. sine of the angle, then divide by two
    d. cosine of the angle

Review Questions

12. The middle ordinate for an eight-point miter is calculated by multiplying the long cutback by _____ degrees.
    a. sine 22½
    b. tangent 45
    c. sine 45
    d. cosine 22½

13. The rule of thumb for building a miter is that the radius should be _____.
    a. half the OD
    b. the same as a long radius elbow of the same OD
    c. the length of the pipe
    d. the same as the radius of a coil

14. A miter in 16-inch pipe would have _____ segments.
    a. two
    b. three
    c. four to seven
    d. ten

15. The number of degrees in a miter cut equals _____.
    a. the number of segments
    b. half the degrees of the full turn
    c. twice the number of segments
    d. the degrees in the turn divided by twice the number of welds

16. The center-to-center length for the segments of a mitered turn is found by _____.
    a. two times sine of miter
    b. radius times sine of miter
    c. tangent of miter
    d. two times radius times tangent of miter

17. The length of the end piece of a mitered turn is _____.
    a. two times the sine of the miter
    b. half the sine of the miter
    c. half the center-to-center length
    d. two times the OD

18. A fishmouth is calculated based on the geometry of _____.
    a. spherical bodies
    b. segments of a circle
    c. conical sections
    d. pyramids

19. A lateral is the intersection of two pipes at _____.
    a. 45 degrees
    b. 90 degrees or more
    c. more than 45 degrees
    d. less than 90 degrees

20. A 6" × 8" 45-degree lateral is to be calculated. What is the number 6 ordinate out of 8?
    a. 0.9936 inches
    b. 3 inches
    c. 6 inches
    d. 12 inches

Summary

Advanced pipe fabrication takes many hours of practice in order to become proficient in the formulas and procedures involved. The various pipefitting handbooks available on the market are good tools to use to increase your skills in advanced pipe fabrication. These handbooks provide useful formulas and techniques to ease the job of fabricating offsets, mitering pipe, laying out laterals, and installing dummy legs. You will find many methods for performing these tasks, and as a pipefitter, you should select the method that produces the most efficient and productive result. Many times, the job site conditions dictate the method that you have to use, and the most skilled pipefitters are those whose knowledge and skills are comprehensive enough to cover a wide range of job site conditions.

Notes

Trade Terms Introduced in This Module

Base line: A straight line drawn around a pipe to be used as a measuring point.

Branch: A line that intersects with another line.

Chord: A straight line crossing a circle that does not pass through the center of the circle.

Cutback: The point at which a miter fitting is to be cut.

Fishmouth: A fabricated 90-degree intersection of pipe.

Ordinate: A division of segments obtained by dividing the circumference of a pipe into equal parts.

Ordinate lines: Straight lines drawn along the length of the pipe connecting the ordinate marks.

Segment: A part of a circle that is defined by a chord and the curve of the circumference.

Resources & Acknowledgments

Additional Resources

This module is intended to present thorough resources for task training. The following reference works are suggested for further study. These are optional materials for continued education rather than for task training.

www.sosmath.com/trig/trig.html

www.analyzemath.com/trigonometry.html

www.counton.org/alevel/pure/purtuttri.htm

NCCER CURRICULA — USER UPDATE

NCCER makes every effort to keep its textbooks up-to-date and free of technical errors. We appreciate your help in this process. If you find an error, a typographical mistake, or an inaccuracy in NCCER's curricula, please fill out this form (or a photocopy), or complete the online form at **www.nccer.org/olf**. Be sure to include the exact module ID number, page number, a detailed description, and your recommended correction. Your input will be brought to the attention of the Authoring Team. Thank you for your assistance.

Instructors – If you have an idea for improving this textbook, or have found that additional materials were necessary to teach this module effectively, please let us know so that we may present your suggestions to the Authoring Team.

NCCER Product Development and Revision
13614 Progress Blvd., Alachua, FL 32615

Email: curriculum@nccer.org
Online: www.nccer.org/olf

❏ Trainee Guide ❏ AIG ❏ Exam ❏ PowerPoints Other _____

Craft / Level:

Copyright Date:

Module ID Number / Title:

Section Number(s):

Description:

Recommended Correction:

Your Name:

Address:

Email:

Phone:

Pipefitting Level Four

08403-07

Stress Relieving and Aligning

08403-07
Stress Relieving and Aligning

Topics to be presented in this module include:

1.0.0	Introduction	3.2
2.0.0	Thermal Expansion	3.2
3.0.0	Performing Stress Relief	3.4
4.0.0	Measuring Temperatures	3.9
5.0.0	Interpass Temperature	3.10
6.0.0	Postheating	3.13
7.0.0	Stress Relief for Aligning Pipe to Rotating Equipment	3.14

Overview

Stress relieving is the process of preheating and postheating to keep welds from distorting a pipe assembly. Alignment is the reason for stress relieving, because if the pipe will not fit up accurately to machinery, dynamically balanced pumps will be unbalanced by the distortions of the piping attachments. This module covers types of misalignment and the techniques for correcting them. Information is provided for preheating and postheating pipe before and after welding, as well as for accurately fitting pipe to pumps and other machinery without inducing pipe stress.

Objectives

When you have completed this module, you will be able to do the following:

1. Explain thermal expansion, anchors, and cold springings.
2. Explain stress-relief procedures.
3. Explain grouting.
4. Explain four types of misalignment.
5. Align pipe flanges to rotating equipment nozzles.

Trade Terms

Alloys
Carbon
Constituents
Critical temperature
Hardenability
Heat-affected zone (HAZ)
Induction heating
Interpass heat
Postheat
Preheat
Preheat weld treatment (PHWT)
Quenching
Stress
Stress relieving
Tempering

Required Trainee Materials

1. Pencil and paper
2. Appropriate personal protective equipment

Prerequisites

Before you begin this module, it is recommended that you successfully complete *Core Curriculum*; *Pipefitting Level One*; *Pipefitting Level Two*; *Pipefitting Level Three*; and *Pipefitting Level Four*, Modules 08401-07 and 08402-07.

This course map shows all of the modules in the fourth level of the Pipefitting curriculum. The suggested training order begins at the bottom and proceeds up. Skill levels increase as you advance on the course map. The local Training Program Sponsor may adjust the training order.

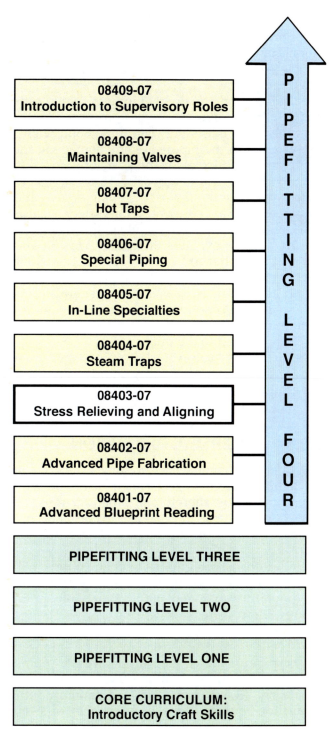

1.0.0 ♦ INTRODUCTION

Most process piping systems are routed between storage tanks and pumps. As a pipefitter, you must be able to properly align piping to pumps and other rotating equipment to avoid excessive stress on the pipe, joints, and rotating equipment. When pipe and rotating equipment are not properly aligned, the stress that exists between these components causes leaks in the piping system, damage to the equipment, and excessive pressure and stress to the pipe hangers and supports. All of these conditions decrease the life of the piping system and equipment and also present serious safety hazards to those who work in the vicinity of the system.

This module explains thermal expansion and the methods that are used to properly relieve stress before aligning pipe to rotating equipment. Pump alignment is explained in general terms only so that the pipefitter will have a better understanding of why it is important that the pipe be correctly aligned to the pump.

Before stress relieving any pipe, you must check your company's policies and procedures governing stress relieving. Some companies do not allow stress relieving of pipe.

2.0.0 ♦ THERMAL EXPANSION

All materials change in size, to some extent, when the temperature changes. This is also true with piping materials; therefore, a pipefitter must be able to account for thermal expansion and contraction while working out the details for the fabrication of a pipeline.

Under normal conditions, metallic piping will change less than an inch in a 100-foot length with a temperature change of 300°F. If only one point in a pipeline were kept fixed during a change in length, movement would take place in perfect freedom, and no stress would be imposed on the pipe or the connection. However, piping systems have more than one fixed point. They are nearly always restrained at terminal points by anchors, guides, stops, rigid hangers, or sway braces. These restraining points resist expansion and put the line under stress, which causes it to deform when the temperature changes.

Stress imposed by expansion must be kept within the strength capability of the pipe and also of the points to which the pipe is anchored. Equipment, such as a pump or a turbine, that is fastened solidly will be put under excessive stress if the pipe fastened to it expands toward it. This must be prevented by allowing expansion away from the equipment. Failure to do so can result in ruptured pipe or damage to the structure or the equipment to which the pipe is anchored.

All materials do not expand and contract at the same rate. This can cause problems when different types of materials are used in the same piping system. *Table 1* shows the expansion of different types of pipe.

To find out how much a length of pipe would expand for a given change in temperature, multiply the length of the pipe in inches, times the temperature change in degrees Fahrenheit, times the coefficient of expansion for that material from *Table 1*. For example, a hundred feet of carbon steel pipe (1,200 inches) that is heated from 100°F to 500°F would expand by (length) 1,200 × (temperature change) 400 × (coefficient of expansion) 0.0000063 = 3.024 inches of expansion.

Charts showing the exact linear expansion for each degree of heat will not always be available to you. However, the following rule of thumb can be used for steel pipe: for all temperatures, allow ¾ inch per 100 feet per 100°F.

2.1.0 Flexibility in Layout

Flexibility can often be built into the layout of a piping system. If a pipeline is built with small-diameter pipe and if it will not have great temperature changes, the system can be designed to allow for expansion. Flexibility in layout consists of making many changes in direction between anchor points. These changes in direction allow for some thermal expansion of the line without damaging the joints and fittings. *Figure 1* shows an example of flexibility in layout of a pipeline constructed of 4-inch, Schedule 40, steel pipe carrying steam indoors at 400°F.

2.2.0 Installing Expansion Loops

The second method used to account for expansion is expansion loops that are installed in the pipe run to provide extra flexibility between anchors. Expansion loops accommodate thermal expansion and contraction and sideways bowing. These loops absorb the movement of the line caused by expansion and contraction. The pipe is anchored along its length and sometimes in the loop itself.

Table 1 Coefficient of Expansion

Material	inches or mm
Carbon steel	0.0000063
Aluminum	0.0000124
Cast iron	0.0000059
Nickel steel	0.0000073
Stainless steel	0.0000095
Concrete	0.0000070

Guides are placed at designated intervals along the pipe run to ensure that the pipe moves toward the expansion loop as it expands.

Expansion loops are made with pipe bends or welded pipe and fittings. They should not be fabricated with flanges or threaded fittings because these types of joints cannot withstand the stress and will leak. Fabricated expansion U-loops made of pipe welded to long-radius elbows are usually very simple in shape, and can be fabricated in the shop according to the specifications on the layout drawing. The number and type of expansion loops used differs with each installation. *Figure 2* shows an expansion loop.

2.3.0 Cold Springing

The third way to deal with expansion is to use cold springing along with one of the other two methods used to allow for expansion. Cold springing is a method of shortening the pipe run when it is installed so that when the pipe heats up, it expands to its intended length. Cold springing also relieves the stress that thermal expansion exerts on the pipe.

The cold-spring gap refers to the difference between the length of the installed pipe and the length that the pipe should be. The location and the size of the cold-spring gap must be calculated accurately and the piping installed according to those calculations. Cold springing pipe at the wrong location or by the wrong amount can damage the pipe joints, anchors, valves, or attached equipment, so it is not advisable to guess at cold-spring dimensions and locations. *Figure 3* shows examples of cold springing.

The most common method of cold springing a pipe assembly is to use a jack to push the pipe out to the point specified to compensate for expansion and weld the pipe there. Expansion will then move the pipe the desired amount.

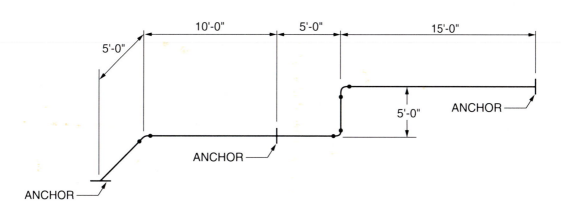

Figure 1 ◆ Flexibility in layout.

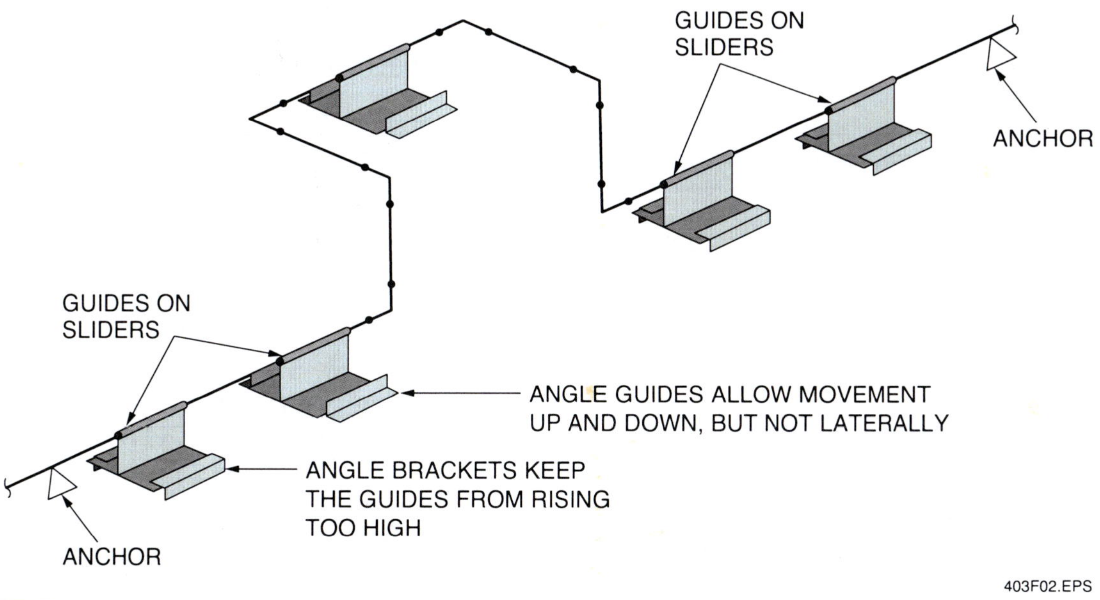

Figure 2 ◆ Expansion loop.

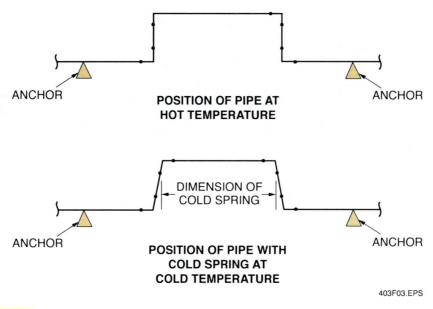

Figure 3 ♦ Cold springing.

3.0.0 ♦ PERFORMING STRESS RELIEF

Pipe stresses are locked into fabricated pipe assemblies by the rapid heating and cooling of pipe that occurs during the welding process. Heat treatment and stress relieving of piping are necessary to relieve stresses created during welding and to prevent stress cracks from forming along the weld. Stresses are caused by temperature differentials between the hot welding zone and the cooler zones. Heat treatment and stress relieving maintain or restore the original crystalline structure of the base metal in the pipe and maintain certain desired properties, such as corrosion resistance.

To stress relieve steel pipe, heat the pipe for about 6 inches on either side of the weld to the temperature that the specifications allow. Maintain this temperature for a period of time according to the size and thickness of the pipe. Cool the pipe slowly and uniformly to room temperature. Uniform cooling prevents development of new stresses, which can equal those created during the welding process.

Time and temperature affect stress relief. Maintaining a lower pipe temperature for a longer period of time achieves the same result as maintaining a higher temperature for a shorter period of time. The heat treatments used in welding piping are **preheating**, **interpass heating**, and **postheating**.

Preheat weld treatment (PHWT) is the controlled heating of the base metal immediately before welding begins. Its main purpose is to keep the weld and the base metal within the **heat-affected zone (HAZ)** along the weld line from cooling too fast. In some cases, the whole structure is preheated. Sometimes, it may not be possible or practical to heat the whole structure. When this is the case, only the section near the weld is preheated.

Interpass temperature control is used to maintain the temperature of the weld zone during welding. It is used when the cooling rate is too high or too low to maintain the correct temperature of the weldment between weld passes. The minimum interpass temperature is usually the same as the preheat temperature. If welding is ever interrupted, the temperature of the weldment must be brought back to the interpass temperature before welding is continued. Usually, there is also a maximum interpass temperature, which is especially important in preventing overheating from the heat buildup in smaller weldments.

3.1.0 Temperature and Metal Structure

At room temperature, most metals are solid with a crystalline structure. They become liquids if heated to a high enough temperature. Temperature is a measure of molecular activity, so when the temperature is high enough, the forces that hold the molecules in their crystalline structures break down and the metal becomes a liquid. At even higher temperatures, metals become gases. The temperature at which a metal becomes liquid is its melting point, and this point is different for every metal. In the case of **alloys** (metal mixtures), the various **constituents** melt at different temperatures. Some, like **carbon**, may not melt at all.

Below the melting point, there is a temperature zone at which crystalline changes take place. Depending on the metal or alloy, these changes can seriously affect the metal's physical structure and mechanical characteristics. Quick cooling (quenching) of a metal while it is in this temperature zone can freeze some of the modified crystalline structure. The result is that the mechanical properties and physical traits of the cooled metal will be very different than if the same metal were cooled more slowly. Quenching does not have much of an effect on low-carbon and mild steels because they do not contain much carbon. However, quenching medium- and high-carbon steels produces a hard and brittle crystalline structure, which reduces the metal's toughness. *Table 2* shows the minimum preheat temperatures for various base materials.

When medium-carbon and high-carbon steel are welded, the temperature in the weld zone always reaches the melting point (above the critical temperature), and the crystalline structure in the weld zone changes. If the metal is not preheated enough, the large mass of the cool base metal will quench the weld zone and cause localized hardness and brittleness. If the metal is adequately preheated, the weld zone cools much more slowly, allowing the crystalline structure to transform back to the crystalline form of the surrounding base metal.

* Martensitic Steel *

Preheating may be required to:

- Reduce shrinkage in the weld and adjacent base metal
- Prevent excessive hardening and reduced ductility of weld filler and base metals in the HAZ
- Reduce hydrogen gas in the weld zone
- Ensure compliance with required welding procedures

3.1.1 Excessive Hardening From Quenching

All welding processes use or generate temperatures higher than the melting point of the base metal. If the welded metal is a high-carbon alloy steel or cast iron that is quenched, a hard, brittle metal called martensite will form in the weld zone. This causes the entire heat-affected zone (HAZ) parallel to the weld fusion line to become hard and brittle (see *Figure 4*). The large temperature difference also causes differential thermal expansion and creates stresses in the weld region. Heavy plate, pipe, and castings have a high heat-absorption capacity, which will cause the weld metal and adjacent base metal to quench unless they are adequately preheated.

Table 2 Minimum Preheat Temperatures for Various Base Materials

BASE MATERIALS	WELDING PROCESSES	MATERIAL THICKNESS	MINIMUM PREHEAT TEMPERATURE (°F)
ASTM A36 ASTM A53 GR. B ASTM A106 GR. B API 5L GR. B API 5L GR. X42	SHIELDED METAL ARC WELDING WITH OTHER THAN LOW HYDROGEN ELECTRODES	LESS THAN OR EQUAL TO ¾"	NONE (SEE NOTE)
		OVER ¾" THROUGH 1½"	150
		OVER 1½" THROUGH 2½"	225
		OVER 2½"	300
ASTM A36 ASTM A53 GR. B ASTM A106 GR. B ASTM A572 GR. 42 ASTM A572 GR. 50 ASTM A586 API 5L GR. B API 5L GR. X42	SHIELDED METAL ARC WELDING WITH LOW HYDROGEN ELECTRODES, SUBMERGED ARC WELDING, GAS METAL ARC WELDING, AND FLUX CORED ARC WELDING	LESS THAN OR EQUAL TO ¾"	NONE (SEE NOTE)
		OVER ¾" THROUGH 1½"	50
		OVER 1½" THROUGH 2½"	150
		OVER 2½"	225
ASTM A572 GR. 60 ASTM A572 GR. 65 ASTM 5L GR. X52	SHIELDED METAL ARC WELDING WITH LOW HYDROGEN ELECTRODES, SUBMERGED ARC WELDING, GAS METAL ARC WELDING, AND FLUX CORED ARC WELDING	LESS THAN OR EQUAL TO ¾"	50
		OVER ¾" THROUGH 1½"	150
		OVER 1½" THROUGH 2½"	225
		OVER 2½"	300
ASTM A514 ASTM A517	SHIELDED METAL ARC WELDING WITH LOW HYDROGEN ELECTRODES, SUBMERGED ARC WELDING, GAS METAL ARC WELDING, AND FLUX CORED ARC WELDING	LESS THAN OR EQUAL TO ¾"	50
		OVER ¾" THROUGH 1½"	125
		OVER 1½" THROUGH 2½"	175
		OVER 2½"	275

NOTE: If below 32°F, preheat to 70°F

3.1.2 Reducing Hydrogen Gas in the Weld Zone

The welding arc can split moisture in the weld zone into hydrogen and oxygen. The hydrogen dissolves in the molten base metal, and the oxygen combines with other elements to form slag. As the metal begins to cool, the hydrogen forms gas bubbles in the base metal along the weld boundary. These bubbles create pressure that strains the metal and causes underbead cracking along the weld boundary and at the toe. Preheating helps remove moisture from the weld zone and results in slower cooling, which allows the hydrogen more time to diffuse up through the filler metal and down into the adjacent base metal. The more time the hydrogen has to diffuse, the less pressure it causes and, therefore, the less chance that underbead cracking will occur. *Figure 5* is a simplified diagram showing underbead cracking caused by hydrogen bubbles.

3.1.3 Complying with Required Welding Procedures

If a site has welding procedure specifications (WPS) or quality specifications for particular welding procedures, it is usually mandatory that they be followed exactly. Check with your supervisor if you are not absolutely sure of the specifications that apply to your task.

3.2.0 Metals That Require Preheating

It can be very difficult to determine whether metal assemblies and conditions require preheating. Preheating and interpass temperature control depend on the base metal composition, thickness, and degree of restraint. Some base metals only require preheating when they exceed a certain thickness or are too cold. For example, metals exposed to freezing winter temperatures usually require some preheating. Complex-shaped welding assembles and castings usually require preheating to avoid severe stresses or warping. Some alloys, such as high-carbon or alloy steels, require preheating to avoid brittleness and cracking. Not all metals require preheating. Low-carbon steels rarely need preheating. *Table 3* shows the metals that usually require preheating.

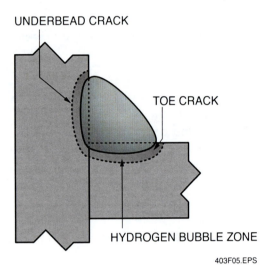

Figure 5 ◆ Underbead cracking caused by hydrogen bubbles.

Table 3 Metals That Usually Require Preheating

Metal or Alloy	Conditions or Forms
Aluminum	Large or thick section castings
Copper	All (prevents too-rapid heat loss)
Bronze, copper-based	All (prevents too-rapid heat loss)
Mild carbon steels	Restricted joints, complex shapes, freezing temperatures, and carbon content over 0.30%
Cast irons	All types and all shapes
Cast steels	Higher carbon content or complex shapes
Low-alloy steels	Thicker sections or restricted joints
Low-alloy steels	Heating temperature dependent on carbon or alloy content
Manganese steels	Heating temperature dependent on carbon or alloy content
Martensitic stainless steels	Carbon content over 0.10%
Ferritic stainless steels	Thick plates

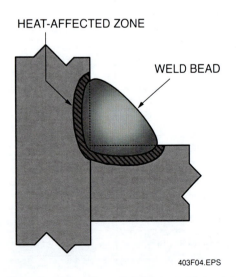

Figure 4 ◆ Heat-affected zone (HAZ).

3.2.1 Preheating Temperatures

Preheating requirements are affected by the alloy composition and thickness of the base metal, the welding process to be used, and the ambient temperature. The hardenability of a steel is directly related to its carbon content and alloying elements. Because different types of steel have different amounts of carbon and alloying elements, they have different preheat requirements. As a general rule, the higher the carbon content, the higher the preheating temperature required. *Table 4* lists preheat requirements for steels.

Temperature-indicating crayons, or temp sticks (*Figure 6*), are often used to measure the preheat temperature of the base metal. They are used by marking the workpiece before heating or by stroking the workpiece with the crayon during heating. When the rated temperature has been reached, a distinct melt or smear will become evident. Other methods to measure preheat temperature will be discussed later in this module.

3.3.0 Preheating Methods

Preheating is most effective when the entire welding assembly is preheated (general heating) in an oven or preheating furnace, because the heating is even. However, because of site conditions or the size or shape of the welding assembly, preheating in this manner is not always possible. Therefore, a variety of heating devices can be used for preheating. Localized heating of specific regions can be done with gas torches, blowtorches, electric resistance heaters, radio frequency induction heaters, or partial ovens that are built only around the region to be heated. Parts of the oven or enclosure can be removed so that the welder can have access and still maintain the preheating temperature.

Most welding shops contain one or more types of preheating devices, which may include:

- Oxyfuel torches
- Portable preheating torches
- Open-top preheaters
- Electric resistance heaters
- Induction heaters

The choice of preheating device is usually specified in the QC requirements for the job. Propane torches are sometimes used when oxyfuel torches are too hot.

3.3.1 Oxyfuel Torches

If properly monitored, oxyfuel torches equipped with specialized heating tips can be used for localized preheating.

NOTE
Because an oxyfuel torch has a localized flame, it is not recommended for preheating large objects.

The specialized heating tips are designed to produce blowtorch flame patterns. Heating tips,

Figure 6 ◆ Temperature-indicating crayon.

Table 4 Preheat Requirements for Steels

Steel Type	Shapes or Conditions	Preheat Temperature
Low-carbon steel (0.10%–0.30% C)	Freezing temperatures	Above 70°F
Low-carbon steel (0.10%–0.30% C)	Complex large shapes	100°F–300°F
Medium-carbon steel (0.30%–0.45% C)	Complex large shapes	300°F–500°F
High-carbon steel (0.45%–0.80% C)	All sizes and shapes	500°F–800°F
Low-alloy steel	Lower C or alloy %	100°F–500°F
Low-alloy steel	Higher C or alloy %	500°F–800°F
Manganese steel	Lower C or alloy %	100°F–200°F
Manganese steel	Higher C or alloy %	200°F–500°F
Cast iron	All shapes	200°F–400°F
Martensitic stainless steel	0.10%–0.20% C	500°F
Martensitic stainless steel	0.20%–0.50% C	500°F and postheat
Ferritic stainless steel	¼" and thicker	200°F–400°F and postheat

like the rosebud-style multiflame heating heads, are designed to mount directly to torch handles like welding tips. Heating tips designed to replace the cutting tips on straight cutting torches or combination cutting torch cutting attachments are also available. *Figure 7* shows different types of heating tips.

> **WARNING!**
> Always make sure that the correct type of tip is employed for the fuel gas being used. Using a tip with the wrong gas can cause the tip to explode. Observe the pressure and flow rates recommended by the manufacturer.

3.3.2 Portable Preheating Torches

Portable preheating torches are designed to burn either oil or gas combined with compressed air. Some use just gas and have an air blower to increase the intensity of the flame. *Figure 8* shows a gas and compressed air torch suitable for preheating large weldments.

3.3.3 Open-Top Preheaters

Open-top preheaters come in several basic designs. The simplest is a large, horizontal gas burner with an adjacent or overhead support structure. The object to be heated is mounted on the support structure directly over the gas burners. A variation is the open flat-top preheater, which consists of a horizontal grate or series of bars with gas burners located underneath (see *Figure 9*). The item to be preheated is placed on the grating. This type of preheater works similarly to a kitchen gas cooking range.

ROSEBUD

HEATING TIP FOR STRAIGHT AND COMBINATION CUTTING TORCHES

Figure 7 ◆ Heating tips.

Figure 8 ◆ Gas preheating torch.

WEEDBURNER

OPEN FLAT-TOP PREHEATER

Figure 9 ◆ Open flat-top preheater.

3.3.4 Electric Resistance Heaters

Electric resistance heaters consist of tubular resistance elements that are formed in several shapes or styles to fit various applications (see *Figure 10*). They are temporarily fastened against the surface of the metal to be heated. Electricity passing through the elements heats the weldment.

3.3.5 Induction Heaters

In induction heating, an alternating magnetic field is generated within the metal body to be heated. A magnetic coil is placed around the metal object, and a strong alternating current is passed through the coil. The resulting alternating magnetic field generated within the base metal creates circulating electrical currents known as eddy currents. The interaction of these currents with the electrical resistance of the base metal produces an electrical heating effect, raising the temperature of the base metal. *Figure 11* shows an air-cooled induction heating system consisting of a power supply, induction blanket, and associated cables. Air-cooled systems are typically used for preheat applications up to 400°F. This type of system may use a controller to monitor and automatically control the temperature. If not equipped with a controller, the use of a temperature indicator is required. The existence of ceramic blanket induction heating systems has made this a desirable option.

4.0.0 ◆ MEASURING TEMPERATURES

Controlling preheating, interpass, and postheating temperatures depends on accurately determining the temperature of the weld zone or assembly. Both underheating and overheating can negatively affect a metal's physical and mechanical characteristics. There are a number of ways to determine the surface temperature of an item being heated. Some involve complex manufactured products, while others use less sophisticated materials.

Devices and materials that can be used to determine surface temperatures include the following:

- Pyrometers
- Thermocouple devices
- Temperature-sensitive indicators

4.1.0 Pyrometers

A pyrometer is a handheld device that measures temperature from the frequencies of the emitted electromagnetic waves (infrared or visible light). Optical pyrometers are non-contact instruments commonly used to measure temperature. They work on the principle of using the human eye to match the brightness of the hot object to the brightness of a calibrated lamp filament inside the instrument. Optical pyrometers like the one shown in *Figure 12* are easy to operate.

The instrument's optics serve as a telescope to provide a clean, enlarged picture of the area to be measured (target). The operator adjusts a control (rheostat) on the instrument while aiming at and viewing the target through the instrument telescope. By this process, the light radiated from the target is compared to the known brightness

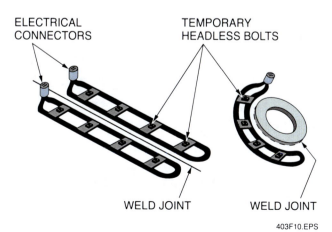

Figure 10 ◆ Resistance heating elements.

Figure 11 ◆ Induction heating system.

of the calibrated lamp inside the instrument. The operator adjusts the control until a color blend is achieved between the radiated target signal and the internal calibrated lamp. The corresponding temperature is then indicated on a direct reading temperature scale. Newer gun-style pyrometer models are available that require the operator only to aim the gun at the target and read the temperature. No adjustments are necessary.

4.2.0 Thermocouple Devices

A thermocouple device is a bimetallic device (usually twisted wires of two dissimilar metals) placed in physical contact with the material whose temperature is to be measured. Thermocouples work on the principle that when two dissimilar metals are joined, a predictable voltage will be generated that relates to the difference in temperature between the measuring junction and the reference junction. Leads from the thermocouple are connected to a device that can measure very small electric voltages and currents. The thermocouple generates a weak electric voltage and current that is in direct proportion to the thermocouple's temperature. The measured voltage or current is fed to a calibrated meter and read directly in degrees of temperature.

4.3.0 Temperature-Sensitive Indicators

Temperature-sensitive indicators are devices that change form or color at specific temperatures or temperature ranges. These include both commercially made materials and devices such as crayon sticks (*Figure 13A*), liquids (*13B*), chalks, powders, pellets, and labels (*13C*) designed to change color or form at precise temperatures, often within a 1 percent tolerance. An example of a commonly used indicator is the temperature stick, mentioned earlier in this module. Temperature sticks are a series of graduated temperature-indicating crayons that cover the temperature range from 100°F to 1,150°F. These sticks are used to make a mark on the base metal. At a specific temperature, the mark melts or changes color. *Table 5* shows a preheating chart and the temperature-indicating crayons to use for each type of metal.

5.0.0 ♦ INTERPASS TEMPERATURE

When preheating is important, interpass temperature for multiple-pass welds is usually also

Figure 12 ♦ Optical pyrometer.

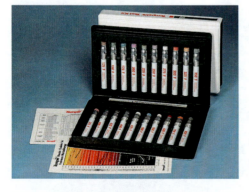

TEMPERATURE STICKS
(A)

TEMPERATURE LIQUID
(B)

TEMPERATURE LABELS
(C)

Figure 13 ♦ Temperature-sensitive indicators.

Table 5 Preheating Chart and Temperature-Indicating Crayons (1 of 2)

METAL GROUP	METAL DESIGNATION	APPROXIMATE COMPOSITION (PERCENTAGE)							RECOMMENDED PREHEAT TEMPERATURE	TEMPERATURE-INDICATING CRAYONS (°F)
		C	MN	BI	CR	NI	MO	CU	OTHER		
PLAIN CARBON STEELS	PLAIN CARBON STEEL	< 0.20								UP TO 150°F	150
	PLAIN CARBON STEEL	0.20–0.30								150°F–300°F	150-200-250-300
	PLAIN CARBON STEEL	0.30–0.45								250°F–450°F	250-300-350-400-450
	PLAIN CARBON STEEL	0.45–0.80								450°F–750°F	450-500-600-700-750
PLAIN MOLY STEELS	CARBON MOLY STEEL	0.10–0.20					0.50			150°F–250°F	150-200-250
	CARBON MOLY STEEL	0.20–0.30					0.50			200°F–400°F	200-250-300-350-400
MANGANESE STEELS	SILICON STRUCTURAL STEEL	0.35	0.80							300°F–500°F	300-400-500
	MEDIUM MANGANESE STEEL	0.20–0.25	1.0-1.75							300°F–500°F	300-400-500
	SAE 1330 STEEL	0.30	1.75							400°F–800°F	400-500-600
	SAE 1340 STEEL	0.40	1.75							500°F–800°F	500-600-700-800
	SAE 1045 STEEL	0.50	1.75							800°F–900°F	600-700-800-900
	12% MANGANESE STEEL	1.25	12.0							USUALLY NOT REQUIRED	
HIGH TENSILE STEELS	MANGANESE MOLY STEEL	0.20	1.65	0.20			0.35			300°F–500°F	300-400-500
	JALTEN STEEL	0.35 MAX	1.50	0.30				0.40		400°F–800°F	400-500-600
	MANTEN STEEL	0.30 MAX	1.35	0.30				0.20		400°F–800°F	400-500-800
	ARMCO HIGH TENSILE STEEL	0.12 MAX				0.50 MIN	0.05 MIN	0.35 MIN		UP TO 200°F	200
	DOUBLE STRENGTH #1 STEEL	0.12 MAX	0.75			0.50–1.25	0.10 MIN	0.50–1.50		300°F–800°F	300-400-500-600
	DOUBLE STRENGTH #1A STEEL	0.30 MAX	0.75			0.50–1.25	0.10 MIN	0.50–1.50		400°F–700°F	400-500-600-700
	MAYARIR STEEL	0.12 MAX	0.75	0.35	0.2-1.0	0.25–0.75		0.80		UP TO 300°F	200-300
	OTISCOLOY STEEL	0.12 MAX	1.25	0.10 MAX	0.10 MAX			0.50 MAX		200°F–400°F	200-300-400
	NAX HIGH TENSILE STEEL	0.15–0.25		0.75	0.60	0.17	0.15 MAX	0.25 MAX	ZR.12	UP TO 300°F	200-300
	CROMANSIL STEEL	0.14 MAX	1.25	0.75	0.50					300°F–400°F	300-400
	A.W. DYN-EL STEEL	0.11–0.14						0.40		UP TO 300°F	200-300
	CORTEN STEEL	0.12 MAX		0.25–1.0	0.5–1.5	0.55 MAX		0.40		200°F–400°F	200-300-400

Table 5 Preheating Chart and Temperature-Indicating Crayons (2 of 2)

METAL GROUP	METAL DESIGNATION	APPROXIMATE COMPOSITION (PERCENTAGE)								RECOMMENDED PREHEAT TEMPERATURE	TEMPERATURE-INDICATING CRAYONS (°F)
		C	MN	BI	CR	NI	MO	CU	OTHER		
HIGH TENSILE STEELS (CONT.)	CHROME COPPER NICKEL STEEL	0.12 MAX	0.75		0.75	0.75		0.55		200°F–400°F	200-300-400
	CHROME MANGANESE STEEL	0.40	0.90		0.40					400°F–600°F	400-500-600
	YOLOY STEEL	0.05–0.35	0.3–1.0			1.75		1.0		200°F–600°F	200-300-400-500-600
	HI-STEEL	0.12 MAX	0.6	0.3 MAX		0.55		0.9–1.25		200°F–500°F	200-300-400-500
NICKEL STEELS	2½% NICKEL STEEL	0.25				2.25				200°F–400°F	200-300-400
	3½% NICKEL STEEL	0.23				3.50				200°F–400°F	200-250-300-350-400
MOLY BEARING CHROMIUM AND NICKEL STEELS	SAE 4140 STEEL	0.40			0.90		0.20			600°F–800°F	600-700-800
	SAE 4340 STEEL	0.40			0.80	1.85	0.25			700°F–900°F	700-800-900
	SAE 4615 STEEL	0.15				1.80	0.25			400°F–800°F	400-500-600
	SAE 4820 STEEL	0.20				3.50	0.25			600°F–800°F	600-700-800
LOW CHROME MOLY STEELS	1¼% CR–½% MO STEEL	0.17 MAX			1.25		0.50			250°F–400°F	250-300-350-400
	2¼% CR–1% MO STEEL	0.15 MAX			2.25		1.0			300°F–500°F	300-400-500
MEDIUM CHROME MOLY STEELS	5% CR–½% MO STEEL	0.15 MAX			5.0		0.5			400°F–800°F	400-500-600
	7% CR–½% MO STEEL	0.15 MAX			7.0		0.5			400°F–800°F	400-500-600
	9% CR–1% MO STEEL	0.15 MAX			9.0		1.0			400°F–800°F	400-500-600
PLAIN HIGH CHROMIUM STEELS	11½–13% CR TYPE 410	0.15 MAX			12.0					400°F–800°F	400-500-800
	16–18% CR TYPE 430	0.12 MAX			17.0					300°F–500°F	300-400-500
	23–27% CR TYPE 448	0.20 MAX			25.0					300°F–500°F	300-400-500
CHROME NICKEL STAINLESS STEELS	18% CR 8% NI TYPE 304	0.07			18.0	8.0				USUALLY DO NOT REQUIRE PREHEAT BUT IT MAY BE DESIRABLE TO REMOVE CHILL	
	25-12 TYPE 309	0.07			25.0	12.0					
	25-20 TYPE 310	0.10			25.0	20.0					
	18-8 CB TYPE 347	0.07			18.0	8.0			CB 10×C		
	18-8 MO TYPE 316	0.07			18.0	8.0	2.5				200
	18-8 MO TYPE 317	0.07			18.0	8.0	3.5				
IRONS	CAST IRON									700°F–900°F	700-800-900
	NI RESIST									500°F–1,000°F	500-700-900-1000
NON FERROUS	NICKEL, MONEL, INCONEL									PREHEAT NOT USUALLY REQUIRED FOR THIN SECTIONS. 300°F–400°F PREHEAT MAY BE DESIRABLE FOR THICK SECTIONS.	
	ALUMINUM-COPPER										

important. Interpass temperature control is often simply a continuation of preheating. Heating equipment and techniques are the same. The interpass temperature is usually the same as the preheat temperature. However, some specifications list both a maximum and a minimum interpass temperature. Sometimes, the interpass temperature may be specified as a maximum temperature. This might be the case with small weldments and frequent weld passes, where too much heat could build up.

With large, high-mass weldments and infrequent weld passes, the base metal can cool below the required preheat temperature if additional heat is not supplied. If the base metal cools below the specified minimum preheat temperature, it must be reheated to the preheat temperature before welding continues.

6.0.0 ◆ POSTHEATING

Postweld heat treatment (PWHT) is the heating of a weldment or assembly after welding is completed. It can be performed on most types of base metals. Postheat equipment and techniques are the same as those used for preheating.

Postheat treatments include the following:

- Stress relieving
- Annealing
- Normalizing
- Tempering

6.1.0 Stress Relieving

Stress relieving is the most commonly used postheat treatment. It is done below the critical (recrystallization) temperature, usually in the 1,050°F to 1,200°F range.

Stress relief treatment is done for the following reasons:

- To reduce residual (shrinkage) stresses in weldments and castings. This is especially important with highly restrained joints.
- To improve resistance to corrosion and caustic embrittlement.
- To improve dimensional stability during machining operations.
- To improve resistance to impact loading and low-temperature failure.

Some codes cover stress relieving. These codes specify the heating rate, holding time, and cooling rate. The heating rate is usually 300°F to 350°F per hour. The holding time is usually one hour for each inch of thickness. The cooling rate is also usually 300°F to 350°F per hour. Stress relieving requires the use of temperature indicators and temperature-control equipment.

As discussed in this module, the most commonly used method for stress relieving weldments is by postweld heat treatment. However, stress relief can also be accomplished mechanically. One method involves attaching a mechanical vibrating device to the weldment during or immediately after welding. When activated, the device vibrates the weldment at a specific resonant frequency. This acts to even the stress distribution within the weldment by means of plastic deformation of the metal's grains. The weight of the weldment generally determines the length of time the weldment is subjected to the vibrations.

6.2.0 Annealing

Annealing relieves stresses much more than a stress relieving postheat treatment can, but it results in a steel of lower strength and higher ductility. Annealing is done at temperatures approximately 100°F above the critical temperature, with a prolonged holding period. The heating is followed by slow cooling in the furnace or by covering, wrapping, or burying the item, although austenitic stainless steel requires rapid cooling. Annealing is used to relieve the residual stresses associated with welds in carbon-molybdenum pipe and also to relieve stresses in welded castings that contain casting strains. Generally, annealing is considered to leave the metal in its softest condition, with good ductility.

6.3.0 Normalizing

Normalizing is the process in which heat is applied to remove strains and reduce grain size. Normalizing is done at temperatures and holding periods comparable to those used in annealing, but the cooling is usually done in still air outside the furnace, and the cooling rate is slightly faster. Normalizing usually results in higher strength and less ductility than annealing. It can be used on mild steel weldments to form a uniform austenite solid solution, to soften the steel, and to make it more ductile. Normalized metal is not as soft and free of stresses as fully annealed metal.

6.4.0 Tempering

Tempering, also called drawing, increases the toughness of quenched steel and helps avoid breakage and failure of heat-treated steel. It reduces both the hardness and the brittleness of hardened steel. Tempering is done at temperatures below the critical temperature, much lower than those used for annealing, normalizing, or stress relieving. Tempering is commonly done to tool steel.

6.5.0 Time-at-Temperature Considerations

In most heat treatment procedures, temperature and time-at-temperature are both specified. Time is very important because metallurgical changes are sluggish at temperatures below the critical temperatures. The maximum temperature is determined by the metal alloy. The holding time at the maximum temperature is based on the metal's thickness, usually one hour for each inch of thickness. The cooling rate is determined by the specific treatment or any applicable code. Temperature control must also be used for postheating. Postheating temperatures are determined in the same manner as preheating temperatures.

7.0.0 ♦ STRESS RELIEF FOR ALIGNING PIPE TO ROTATING EQUIPMENT

The method of dry washing welds is used to perform the final alignment of pipe to rotating equipment, such as pumps that are driven by electric motors. If the pipe flange is not properly aligned to equipment flanges, excessive stresses are imposed on the equipment and the pipeline. You must be able to make the connection between the pipe flange and equipment nozzle without forcing the pieces together. If you have to use force to align the flange and equipment nozzle, the pipe will pull the pump and motor out of alignment and decrease the lives of both. Dry washing welds is another form of stress relieving for pipe alignment. Dry washing welds involves rewelding a weld without adding more filler metal to the weld in order to straighten a welded pipe in the desired direction. If the pipe needs to be moved a large distance, you can grind the weld cap off the weld and then reweld, adding filler metal.

Pipe alignment to rotating equipment takes coordination of work activities between the pipefitter installing the pipe and the millwright installing the pump and motor. Before final pipe alignment can be made, the equipment must be set and grouted in place by the millwright. The millwright will normally perform a preliminary alignment between the pump and motor before the final alignment with the pipelines installed.

7.1.0 Grouting

The two basic types of grout are cementitious and epoxy. Check job specifications to determine the appropriate type of grout. Cementitious grout is cheaper, but epoxy grout should be used in applications where the equipment foundation is subjected to chemical contamination or corrosion.

The three primary reasons for grouting under the bases of machinery are:

- To spread the impact of equipment operation over a larger area
- To provide higher compressive strength than reinforced concrete
- To give a more uniform support by filling in all the voids along the base of the equipment

Grout is also used for the installation of baseplates, anchor bolts, bearing plates, and other types of fasteners. Grout must be confined vertically until the drying process is complete. Equipment that is resting on isolators should not be grouted because the grouting defeats the purpose of the isolators.

A concrete mixture should never be used in place of grouting. Grout expands as it sets up, thus filling voids under the bases and bedplates. Filling the voids provides more uniform distribution of the equipment weight and therefore makes the equipment alignment easier to maintain. In contrast, concrete shrinks as it dries, making alignment more difficult to maintain.

Rodding and working the grout is the most common way to apply it. The forms used should be 6 inches wide on the side where the grout is to be poured with 2-inch sides elsewhere. The form should be high enough to force the grout to flow under the equipment. If not used too vigorously, a piece of metal strapping can be used to work the grout under the equipment. Quality grouting depends on adequate flowability and quick placement of the grout. If possible, all vibration should be eliminated where the grout is to be poured because vibration causes the grout to set up improperly, thus weakening the grout.

7.2.0 Performing Preliminary Alignment Between Equipment

Rotating shafts are usually aligned with the coupling halves in place. The measurements are taken from the coupling halves, and adjustments are made accordingly. When aligning two coupling halves so that the shafts will be aligned, there are two ways that they must be lined up. First, the outer diameters, or rims, of the couplings must be lined up all the way around; then the faces of the couplings must be lined up. For this reason, conventional alignment is also called rim-and-face alignment. Although there are only two ways that a coupling can be misaligned, there are four ways of looking at the misalignment.

The four basic types of misalignment are as follows:

- Vertical offset
- Horizontal offset
- Vertical angularity
- Horizontal angularity

7.2.1 Vertical Offset

Vertical offset is also called parallel misalignment. Vertical offset occurs when one of the coupling halves is not in line with the other when viewed from the side. In vertical offset, one coupling is higher than the other. In this type of misalignment, one of the units must be raised or lowered to align the couplings. *Figure 14* shows vertical offset.

7.2.2 Horizontal Offset

Horizontal offset occurs when the outer diameters of the couplings are misaligned from side to side as viewed from the top. In this type of misalignment, one of the units must be moved to one side or the other to align the couplings. *Figure 15* shows horizontal offset.

7.2.3 Vertical Angularity

Vertical angularity is also called angular misalignment. Vertical angularity means that the faces of the couplings are not square with one another when viewed from the side. In this type of misalignment, one of the units must be tilted on its base to align the couplings. *Figure 16* shows vertical angularity.

7.2.4 Horizontal Angularity

Horizontal angularity means that the faces of the couplings are not square with one another when viewed from the top. In this type of misalignment, one of the units must be rotated on its base to align the couplings. *Figure 17* shows horizontal angularity.

Although these are the four basic types of misalignment, any variation of the four types may occur. You may have horizontal offset and vertical angularity, or you could have a misalignment problem that includes all four types.

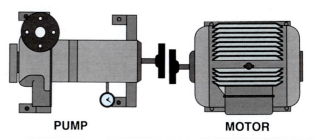

Figure 15 ◆ Horizontal offset.

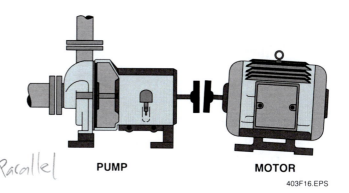

Figure 16 ◆ Vertical angularity.

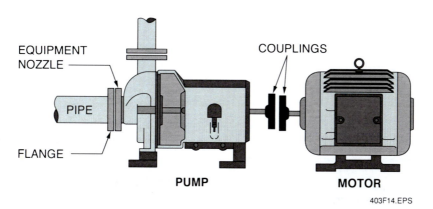

Figure 14 ◆ Vertical offset.

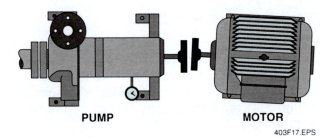

Figure 17 ♦ Horizontal angularity.

7.3.0 Aligning Pipe Flanges to Equipment Nozzles

As a pipefitter, your main concern is to connect the piping system to the rotating equipment without imposing any stress on the equipment. In order to do this, the pipe flange must be properly aligned with the equipment nozzle. Never force a pipe flange and equipment nozzle together. If the flange and nozzle are out of line, you must move the flange into alignment with the nozzle by either heating a concentrated area of the pipe to draw the pipe or by dry washing a weld.

7.3.1 Dry Washing Welds

Follow these steps to dry wash a weld:

Step 1 Identify the first pipe spool next to the equipment.

Step 2 Verify the equipment location coordinates and elevation.

Step 3 Support the pipe adequately in position.

Step 4 Ensure that all welding on the spool is complete.

Step 5 Set the spring hangers to the cold setting.

Step 6 Perform a preliminary stress alignment check.

Step 7 Perform all testing on the pipe.

Step 8 Fit the pipe in the proper location to determine whether stress relief is needed.

Step 9 Fit the pipe in the proper location to minimize stress.

Step 10 Determine the allowable tolerance for stress of the pipe to the equipment.

Step 11 Protect all openings on equipment.

> **NOTE**
> Maintain purge while stress relieving stainless or alloy pipe.

Step 12 Check with the millwright for final pipe alignment to equipment to ensure the proper gap between the pipe and pump.

Step 13 Determine tight spots between the flange and equipment, using a feeler gauge.

Step 14 Determine in which direction the pipe needs to be drawn.

Step 15 Mark the point where the flanges are closest to the area of the cap weld that is to be dry washed.

Step 16 Reweld the cap weld without adding filler metal.

Step 17 Control cooling, using accepted standards.

> **CAUTION**
> Avoid excessive cooling caused by adding too much water.

Step 18 Check for proper alignment of the pipe flange to the equipment flange that is to be connected.

Step 19 Repeat Steps 13 through 16 until proper alignment is achieved. More drawing of the pipe can be accomplished by grinding the cap down and rewelding.

Step 20 Install the proper gasket between the flanges.

Step 21 Install the proper bolts in the flanges.

> **NOTE**
> The following steps should be performed while the millwright monitors the dial indicators located between the pump and motor shafts.

Step 22 Use the crossover method of tightening the bolts, and only tighten the bolts to 20 percent of the recommended torque at a time. If the dial indicators move more than 0.002 inch while tightening the flange bolts, remove the flange bolts and gasket and realign the flanges. If the dial indicators do not move more than 0.002 inch throughout the entire tightening sequence, the flanges are properly aligned.

Step 23 Recheck the dial indicator reading after the final tightening of the flange bolts.

Review Questions

1. When temperature changes, all materials change in _____.
   a. color
   b. light absorption
   c. size
   d. texture

2. To determine the expansion of a given piece of material, multiply the length of the pipe times the temperature change in degrees Fahrenheit times the _____ for that material.
   a. coefficient of expansion
   b. modulus of elasticity
   c. crystallization
   d. diameter of the pipe

3. Flexibility in layout consists of making many _____ between anchor points.
   a. secondary anchors
   b. hangers
   c. changes in direction
   d. pipe breaks

4. Expansion loops should *not* be fabricated with _____.
   a. socket weld fittings
   b. flanged or threaded joints
   c. buttweld joints
   d. mitered joints

5. Pipe stresses are locked into fabricated pipe assemblies by the _____ that occurs in the welding process.
   a. alignment change
   b. welding gap
   c. rapid heating and cooling
   d. uniform cooling

6. The main purpose of PHWT is to keep the weld and the base metal from _____.
   a. heating too fast
   b. cooling too fast
   c. melting
   d. getting too hot

7. Quenching medium- and high-carbon steels reduces the _____.
   a. diameter of the pipe
   b. toughness of the metal
   c. weldability of the metal
   d. length of cooling required

8. When medium- and high-carbon steels are *not* preheated enough, the large mass of base metal will _____ the weld zone.
   a. bend
   b. quench
   c. melt with
   d. crack

9. In induction heating, _____ is generated within the metal.
   a. DC current
   b. AC current
   c. an alternating magnetic field
   d. liquid coolant

10. Offset misalignment is also called _____ misalignment.
    a. shaft
    b. parallel
    c. face
    d. angular

Summary

The fact that piping systems frequently connect dynamic (moving) equipment to static (nonmoving) equipment complicates the pipefitter's task. The process of alignment is further complicated by the temperature differential between live processes and ambient temperature. Pipe expands and contracts depending on the temperature. This complicates the task of aligning pipe to pumps and other process equipment. Misalignment under any circumstance produces leaks, stresses, and possible damage to equipment, as well as hazards to people. When the pipefitter comes in after the equipment is set, and is required to align the pipe with the equipment, precise alignment is the responsibility of the pipefitter.

Misalignment can be offset or angular, depending on whether the flange or coupling faces are parallel or not. When aligning pipe to rotating equipment, you must communicate with the millwrights and the other craftworkers you are working with to plan your activities and complete the required tie-ins to the equipment.

Notes

Trade Terms Introduced in This Module

Alloy: A metal that has had other elements added to it, which substantially changes its mechanical properties.

Carbon: An element which, when combined with iron, forms various kinds of steel. In steel, it is the carbon content that affects the physical properties of the steel.

Constituents: The elements and compounds, such as metal oxides, that make up a mixture or alloy.

Critical temperature: The temperature at which iron crystals in a ferrous-based metal transform from being face-centered to body-centered. This dramatically changes the strength, hardness, and ductility of the metal.

Hardenability: A characteristic of a metal that makes it able to become hard, usually through heat treatment.

Heat-affected zone (HAZ): The area of the base metal that is not melted but whose mechanical properties have been altered by the heat of the welding.

Induction heating: The heating of a conducting material by means of circulating electrical currents induced by an externally applied alternating magnetic field.

Interpass heat: The temperature to which a metal is heated while an operation is performed on the metal.

Postheat: The temperature to which a metal is heated after an operation is performed on the metal.

Preheat: The temperature to which a metal is heated before an operation is performed on the metal.

Preheat weld treatment (PHWT): The controlled heating of the base metal immediately before welding begins.

Quenching: Rapid cooling of a hot metal using air, water, or oil.

Stress: The load imposed on an object.

Stress relieving: The even heating of a structure to a temperature below the critical temperature followed by a slow, even cooling.

Tempering: Increases the toughness of quenched steel and helps avoid breakage and failure of heat-treated steel. Also called *drawing*.

Resources & Acknowledgments

Additional Resources

This module is intended to be a thorough resource for task training. The following reference work is suggested for further study. This is optional material for continued education rather than for task training.

Welding Trainee Guide, Contren Series, Prentice Hall, 2003.

Figure Credits

Topaz Publications, Inc., 403F06, 403F09 (photo)
Thermadyne, Inc., 403F07
Belchfire Corp., 403F08
Miller Electric Mfg. Co., 403F11
Pyrometer Instrument Company, Inc., 403F12
Tempil, Inc., 403F13

NCCER CURRICULA — USER UPDATE

NCCER makes every effort to keep its textbooks up-to-date and free of technical errors. We appreciate your help in this process. If you find an error, a typographical mistake, or an inaccuracy in NCCER's curricula, please fill out this form (or a photocopy), or complete the online form at **www.nccer.org/olf**. Be sure to include the exact module ID number, page number, a detailed description, and your recommended correction. Your input will be brought to the attention of the Authoring Team. Thank you for your assistance.

Instructors – If you have an idea for improving this textbook, or have found that additional materials were necessary to teach this module effectively, please let us know so that we may present your suggestions to the Authoring Team.

NCCER Product Development and Revision
13614 Progress Blvd., Alachua, FL 32615

Email: curriculum@nccer.org
Online: www.nccer.org/olf

❏ Trainee Guide ❏ AIG ❏ Exam ❏ PowerPoints Other _____

Craft / Level: Copyright Date:

Module ID Number / Title:

Section Number(s):

Description:

_____

_____

_____

_____

Recommended Correction:

_____

_____

_____

_____

Your Name:

Address:

_____

Email: Phone:

Pipefitting Level Four

08404-07

Steam Traps

08404-07
Steam Traps

Topics to be presented in this module include:

1.0.0	Introduction	4.2
2.0.0	Steam Traps	4.2
3.0.0	Steam Trap Installation	4.5
4.0.0	Troubleshooting Steam Traps	4.8

Overview

Steam piping systems are subject to condensation, where the steam turns into liquid water at various low points in the piping. The steam can pick the liquid up and project it at high speed into bends and fittings, a phenomenon known as water hammer. So much energy is involved that it can damage pipe and fittings. Low points in steam systems are protected with steam traps, which allow the steam to pass through and let the liquid and air escape the system without allowing steam loss. There are several varieties of steam traps, including float thermostatic, inverted bucket, orifice, thermodynamic, and balanced pressure. This module examines the different types and characteristics of steam traps, as well as maintenance issues.

Objectives

When you have completed this module, you will be able to do the following:

1. Identify types of steam traps.
2. Install steam traps.
3. Troubleshoot steam trap systems.

Trade Terms

Condensate
Drip leg
Flash steam
Superheated steam
Tracer
Water hammer

Required Trainee Materials

1. Pencil and paper
2. Appropriate personal protective equipment

Prerequisites

Before you begin this module, it is recommended that you successfully complete *Core Curriculum*; *Pipefitting Level One*; *Pipefitting Level Two*; *Pipefitting Level Three*; and *Pipefitting Level Four*, Modules 08401-07 through 08403-07.

This course map shows all of the modules in the fourth level of the Pipefitting curriculum. The suggested training order begins at the bottom and proceeds up. Skill levels increase as you advance on the course map. The local Training Program Sponsor may adjust the training order.

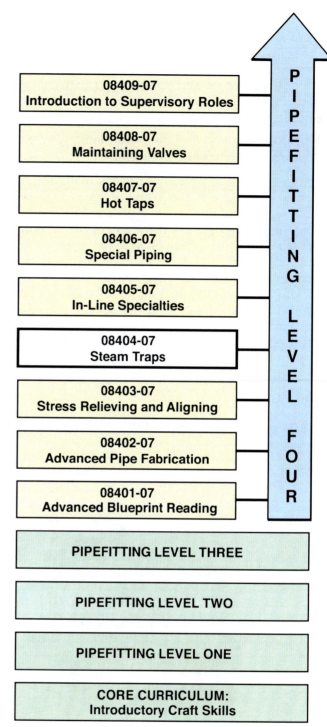

1.0.0 ♦ INTRODUCTION

Steam is used in process plants because it is an efficient way to supply large quantities of heat. For steam to remain efficient, it must be dry and of the proper temperature and pressure. A steam trap is a self-actuating drain valve that does the following:

- Removes air or gases and condensate
- Responds to load and pressure changes
- Closes before passing live steam

A steam trap essentially operates like a toilet. Its main purpose is to flush out the liquid condensate. Traps maintain high energy-transfer rates by constantly purging a steam system of condensate and gases. Air and carbon dioxide can cause problems in a steam line. Air is an excellent insulator and can reduce the efficiency of the steam at its working point on the heat transfer surfaces. Carbon dioxide can dissolve in the condensate and form carbonic acid, which is extremely corrosive. Scale and debris are also transferred with the condensate to areas that are difficult to repair when clogged. A well-placed strainer and steam trap of the proper type can prevent these problems.

2.0.0 ♦ STEAM TRAPS

There are three general types of steam traps: mechanical, thermostatic, and thermodynamic. Among the many variations of these are also the various materials used, such as cast iron, stainless steel, forged steel, and cast steel. Steam traps are fitted to the pipe with screwed, flanged, or socket weld fittings.

It is important to have the correct trap for the application. A steam trap of the right size and type will operate reliably, prevent steam loss, and maximize product output. The wrong trap will wear out prematurely and fail to drain the condensate.

All types of steam traps automatically open an orifice, drain the condensate, then close before steam is lost. With the exception of the inverted bucket-type, no steam trap releases any steam when new or in good condition.

2.1.0 Mechanical Steam Traps

Mechanical steam traps respond to the difference in density between steam and condensate. The trap opens to condensate and closes to steam. The first type of trap was the orifice trap, a mechanical trap in which a small orifice allowed condensate to escape. The typical mechanical trap is called an inverted bucket trap (*Figure 1*) and has only two moving parts: the valve lever assembly and the bucket. The trap is normally installed between the steam source and the condensate return header.

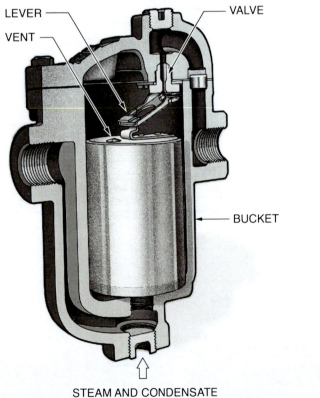

Figure 1 ♦ Inverted bucket trap.

The cycle starts with the bucket down and the valve all the way open. Gases and air flow through first, then steam raises the bucket and closes the valve. The gas slowly bleeds out through a small vent in the top of the bucket. In operation the condensate fills the body and the bucket, causing the bucket to sink and open the discharge valve. The valve remains open until enough steam has collected to float the bucket again.

Since the inverted bucket trap operates on pressure differential, it must be primed before being put into service. The inverted trap can handle high-pressure steam, resists damage from pressure surges and water hammer, and tolerates freezing if the body is made of a ductile material. Its disadvantages are its limited air discharge capacity and a tendency to be noisy.

2.2.0 Thermostatic Steam Traps

As the name implies, thermostatic steam traps respond to temperature changes in the steam line. They open when cooler condensate is present and close to the higher steam temperatures. The following are four types of thermostatic steam traps:

- Liquid expansion
- Balanced pressure
- Bimetallic
- Float

Figure 2 shows the types of thermostatic steam traps.

The liquid expansion thermostatic trap remains closed until the condensate cools below 212°F, opening the trap. This discharges only cool condensate at a constant temperature, which improves heat recovery from submerged tank coils and tracer lines.

The balanced pressure trap opens when the condensate cools slightly below the steam saturation temperature at any pressure within the trap's range. Controlled by a liquid-filled bellows, the discharge valve closes when hot condensate vaporizes the liquid in the bellows, and opens when the condensate cools enough to lower the pressure inside the element. Balanced pressure traps are smaller than mechanical traps, open wide when cold, readily purge gases, and are unlikely to freeze.

The bimetallic thermostatic trap has bimetallic strips that respond to cool condensate by bending to open the condensate valve. When the steam hits the strips, they expand and close the valve. They require considerable cooling to open back up, and the pressure-compensating characteristics vary among the models. The closing force of some designs varies with the steam pressure, approximating the response of a balanced pressure trap. This trap has a large capacity but responds slowly to process load changes. It vents gases well; is not easily damaged by freezing, water hammer, or corrosion; and handles the high temperatures of superheated steam.

The float thermostatic trap has a float that rises when condensate enters and opens the main valve that lies below the water level. At the same time, another discharge valve at the top of the trap, operated by a thermostatic bellows, opens to release cooler gases. This trap vents gases well and responds to wide and sudden pressure changes, discharging condensate continuously. In the failure mode, the main valve is closed but the air vent usually is open. Float thermostatic traps should always be protected from freezing since they usually contain water.

2.3.0 Thermodynamic Steam Traps

Thermodynamic steam traps (*Figure 3*) use the heat energy in hot condensate and steam to control the opening and closing of the trap. They operate well at higher pressures and can be installed in any position because their function does not depend on gravity. They are most often used to drain tracer lines that are in the pressure range of 15 to 200 psig.

The thermodynamic disc steam trap is the most widely used. In this unit, the only moving part is the disc. Flash steam pressure from hot condensate keeps the disc closed. When the flash steam above the disc is condensed by cooler condensate, the disc is pushed up and remains open until hot condensate flashes again to build up pressure in the cap chamber. At the same time, flashing condensate on the underside of the disc is discharged at high velocity, lowering the pressure in this area and slamming the disc shut before steam can escape.

The thermodynamic steam trap can withstand freezing, high pressure, superheating, and water hammer without damage. The audible clicking of the disc indicates when the discharge cycle occurs. These traps last longer if they are not oversized for the load.

2.4.0 Strainers

In all systems, a strainer should be installed upstream from the steam trap. The scale and corrosion in any steam system has to be stopped before it enters the trap, or it will clog and damage the steam trap. Strainers need to be cleaned and inspected on a regular basis. Strainers are available in many styles and are sometimes incorporated into the steam trap. *Figure 4* shows three types of strainers.

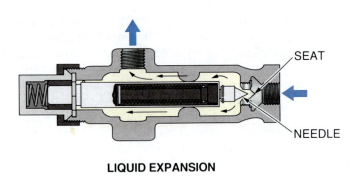
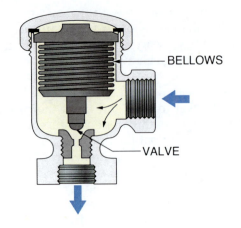
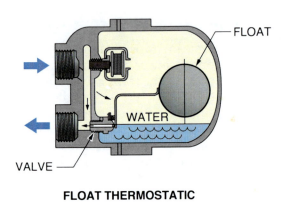
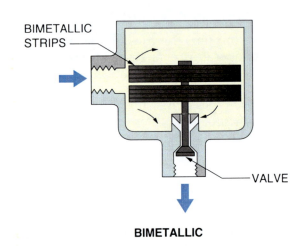

Figure 2 ◆ Thermostatic steam traps.

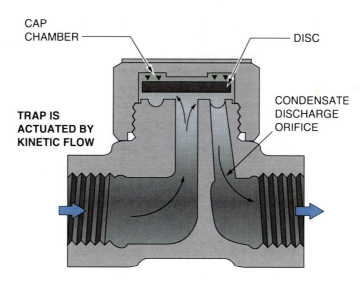

Figure 3 ◆ Thermodynamic steam trap.

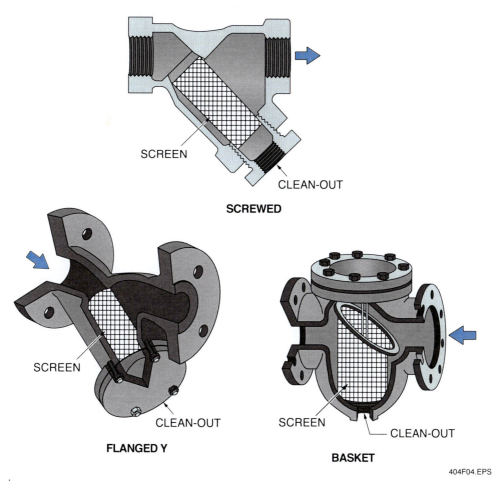

Figure 4 ◆ Strainers.

3.0.0 ◆ STEAM TRAP INSTALLATION

To mount a steam trap anywhere other than the correct position would negate the use of the steam trap. Two general rules for steam trap mountings are that they should be lower than any line in the system and that a strainer has to be upstream of the trap. Isolation drains, couplings, check valves, and drip leg placement are also important to each application, and their use depends on the needs and contents of the system. Each steam trap manufacturer has resource books that detail how to install steam trap systems. Some of these are referenced at the back of this module.

The following is a list of general guidelines for proper steam trap installation:

- Provide a separate trap for each piece of equipment or apparatus. Short-circuiting may occur if more than one piece of equipment is connected to a single trap.
- Tap the steam supply off the top of the steam header to obtain dry steam and avoid steam line condensate.
- Install a supply valve close to the steam header to allow maintenance or revisions to be performed without shutting down the steam header.
- Install a steam supply valve close to the equipment entrance to allow equipment maintenance work to be performed without shutting down the supply line.
- Connect the condensate discharge line to the lowest point in the equipment to avoid water pockets and water hammer.
- Install a shutoff valve upstream of the condensate discharge piping to cut off discharge of condensate from equipment and allow service work to be performed.
- Install a strainer and strainer flush valve ahead of the trap to keep rust, dirt, and scale out of working parts and to allow blow-down removal of foreign material from the strainer basket.
- Provide unions on both sides of the trap for its removal or replacement.
- Install a test valve downstream from the trap to allow observation of discharge when testing.
- Install a check valve downstream from the trap to prevent condensate flow-back during shutdown or in the event of unusual conditions.
- Install a downstream shutoff valve to cut off equipment condensate piping from the main condensate system for maintenance or service work.
- If a bypass around the trap is installed, trap replacement can occur without loss of flow.

A steam trap should have a provision for inspection without interrupting the process flow. The gate valves as arranged in the piping schematic show a generic steam trap and strainer layout. *Figure 5* shows basic trap installation.

The biggest use of traps in a steam system is to clear the distribution headers of condensate. Headers are sometimes sloped to a drip leg that is drained by the trap. A low-capacity trap that operates near the temperature of saturated steam is adequate and operates under varying steam pressures.

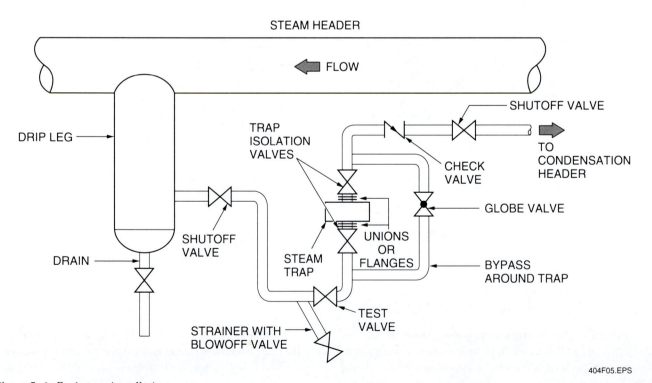

Figure 5 ◆ Basic trap installation.

Figure 6 shows a steam header draining system.

Process heating applications call for a steam trap system that vents air well, withstands outside temperature extremes, continuously discharges hot condensate, responds quickly to load changes, and is unaffected by changing inlet pressure. The float thermostatic steam trap in series with a liquid expansion steam trap works well in these conditions. *Figure 7* shows an outdoor process trap system.

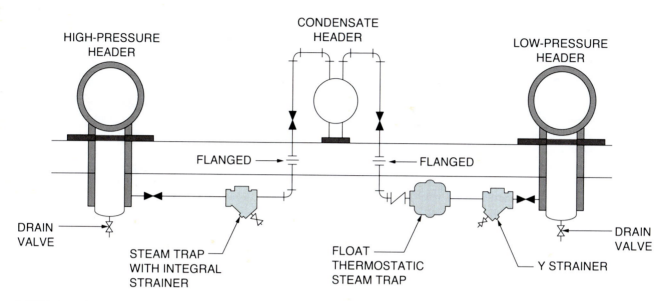

NOTE: Because of the pressure involved, the connection of high and low pressure headers to the condensate header would have a flanged joint.

Figure 6 ♦ Steam header draining system.

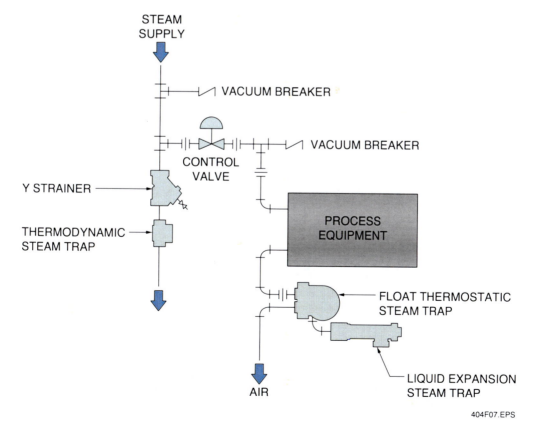

Figure 7 ♦ Outdoor process trap system.

NOTE

Remember that many process systems require piping to slope toward one part of the system or another, so that drainage in the absence of pressure will occur in the correct direction.

4.0.0 ♦ TROUBLESHOOTING STEAM TRAPS

To diagnose and solve problems with steam traps requires listening to the noises they make and measuring temperature and pressure. The temperature of steam can be measured with handheld or remote-sensing pyrometers that read in degrees centigrade or Fahrenheit. Steam is produced at 212°F when at atmospheric pressure. When the steam systems are functioning, the temperature will be substantially higher, as it is confined inside the system, and the pressure is higher as well. Pressure readings are in two scales, absolute (psi) and gauge (psig). Absolute pressure is read from zero up, which is to say that normal atmospheric pressure is 14.7 psi at sea level. Gauge pressure is relative to atmospheric pressure at sea level; therefore, zero psig is equal to 14.7 psi absolute. Pressures below zero gauge are expressed in inches of mercury. Steam trap manufacturers have excellent books and web documents that detail how to troubleshoot and repair steam traps. Some of these are referenced at the back of this module.

4.1.0 Diagnostic Methods

Three basic diagnostic methods of reading a steam system are sight, sound, and temperature. The criteria observed include pressure. The other process, less used to date, is a conductance probe, which acts by comparing the difference in electrical conductivity between condensate and live steam.

Each of the three main methods has advantages and limitations. Sight is fairly immediate; it is easy to tell whether there is condensate coming from a trap. If the condensate line is open, or if a test valve has been added on the condensate line, the condensate can be vented to atmosphere, and it will be obvious that the trap is removing condensate from the line. If live steam is going into the condensate line, the trap has failed open, and it is necessary to resolve the problem. If no condensate is coming from the line, the trap may have failed closed. The limitation of sight diagnosis is that the worker must be able to tell the difference between flash steam and live steam, and it is necessary for the worker to check by activating a test valve. It is also possible for the trap to be overloaded or loaded from the return system, producing an apparent open failure.

The two most commonly used diagnostic tools are temperature sensing and sound, either at human hearing ranges or ultrasound. Note that temperature and sound together are a much more sensitive test than one or the other alone.

From a diagnostic point of view, there are two general types of traps. Traps either flow or dribble continuously or intermittently (on and off). Float thermostat (F and T) traps discharge pretty continuously, if they are working properly. Inverted bucket, bellows thermostatic, thermodynamic, and most thermostatic traps flow intermittently. This allows the person checking the trap to visually determine if the trap is failing, in the more extreme cases, although the generality is not an absolute solution, because a partial failure or overloading may cause misinterpretation.

A few common characteristic symptoms of traps when tested might include the following:

- A thermal analysis will show the inlet as a high temperature area for any of the traps. In most functioning traps, the condensate outlet, whether it is a closed return or an open return, will be cooler than the inlet. Remember that the condensate will most frequently be at steam temperature, but it should be lower than the inlet. The live steam outlet should be hotter than the condensate outlet. If this is not the case, the probability is that the trap is malfunctioning.

- Normal failure for an inverted bucket trap is open failure. The trap loses its prime, and steam and condensate blow through steadily. If this happens, the continuous rushing sound of the blowthrough may be accompanied by the bucket banging against the inside of the trap. The temperature of the trap and of the condensate line will rise, since it is no longer holding condensate. The sound of the bucket linkage rattling inside the trap may indicate that the trap is beginning to loosen up and needs checking. With ultrasound equipment the inverted bucket or thermodynamic traps would show a cyclic curve on the screen in normal function.

- Normal failure for a thermodynamic trap is open; that is, passing steam. Usually, the disk clicks shut audibly once for each cycle. When the disk is cycling normally, it will shut about 4 to 10 times a minute. When the trap fails, the disk no longer clicks as it rises and falls, and steam blows through. If the disk produces a continuous, rapid, rattling sound, the disk is worn, and more problems will probably develop. If a continuous hissing sound is produced, the trap is not cycling. Again, the condensate line will heat up, from escaping steam and the absence of cooler condensate.

4.8 PIPEFITTING ♦ LEVEL FOUR

- Bellows or bimetallic thermostatic traps operate on a difference in temperature between condensate and steam. Both types are intermittent in operation. When there is little condensate, they stay closed most of the time. When there is a lot of condensate, for example at startup, they may run continuously for a long time. In the case of bimetallic traps, misalignment may allow steam leaks. With sound testing equipment, a constant rushing sound in the closed part of the cycle would indicate a leak.
- Float thermostatic traps tend to fail closed. A leaky float will not float on the condensate, or the float may have been crushed by water hammer. In either case, the F and T trap will remain closed and not cycle. The trap will remain cool, and a temperature sensor will show that fact. With listening equipment, a failed float will be silent, in the absence of discharging condensate.
- The thermostatic element in an F and T trap is normally quiet; a rushing sound would indicate that the element had failed open. A rattling or metallic clanging noise might mean that the mechanical linkage had sustained some damage.

Figure 8 ♦ Ultrasonic tester.

The sight method uses a test valve that vents the process steam to the atmosphere for visual inspection. This test is subjective and depends on skill and experience. Since most traps are dealing with condensate that is at steam temperatures, that is, somewhere above the boiling point of water, when condensate is released from the pressure of the steam system in the trap or in a condensate release, a certain amount of the condensate will immediately turn into flash steam. The difference is fairly visible once you have seen the difference a few times; the live steam comes out at first as a hard, blue-white straight flow, often with a clear bluish area at the outlet. Flash steam usually billows and spreads more quickly, as much of the pressure is dissipated in the initial release.

The sound probes are either audible-range or ultrasonic-range probes that pick up noise from the steam system and send an audible signal to headphones. The technician performing this test must also have skill and experience at interpreting the noise as leaks, discharges, or other steam trap problems. *Figure 8* shows a portable ultrasonic tester.

Temperature-sensing tests use pyrometers (*Figure 9*) that are handheld for external readings, or mounted thermowells (*Figure 10*) that read actual steam temperature. A thermowell is a permanently installed well or cavity in a process pipe or tank into which a glass thermometer or thermocouple can be inserted.

The input from these and other test methods must be applied to problems and causes of problems in the steam trap system.

Figure 9 ♦ Pyrometer.

Figure 10 ♦ Thermowells.

4.2.0 Maintaining Steam Traps

Most steam traps fail in the open position, which is difficult to pinpoint because it does not affect equipment operation. When the trap fails open, both steam and condensate flow freely from the steam header to the condensate header, keeping the heat transfer high. The traps that fail in the closed position are easy to identify because the backup of condensate cools the system.

An effective preventive maintenance (PM) program includes scheduled checks of the entire system. Clogged strainers, leaking joints, leaking valve packing, or missing insulation are some of the PM items to check. When the internal parts of a steam trap wear out, the water seal deteriorates, steam flows through the valve assembly, and this erodes the seal into a worsening condition.

Common causes of steam trap failure are:

- Scale, rust, or corrosive buildup preventing the valve from closing
- Valve assembly wear
- Defective or damaged valve seat
- Physical damage from severe water hammer
- Foreign material lodged between seat and valve
- Blocked, clogged, or damaged strainers
- Increased back pressure

Other failures are specific to the type of steam trap. The two indications that there is a failure in a steam trap are that the trap blows steam and the trap will not pass condensate. The following sections explain the causes and solutions for failures on each type of steam trap.

4.2.1 Inverted Bucket Trap

If an inverted bucket trap blows steam, check for loss of the water seal. Isolate the trap; wait for condensate to accumulate, and start the steam flow again. If this solves the problem, try to discover the cause of the water seal loss. It could be caused by superheat, sudden pressure fluctuations, or the trap being installed so that the water seal runs out by gravity.

Another solution is to try installing a check valve before the trap. If the steam blow persists, check for dirt or wear on the valve and linkage. Replace the valve, the seat, and the lever.

If this fails, check the bucket. If the bucket or lever is distorted, it was probably caused by water hammer. Trace the source of the problem and try to eliminate it.

If the trap will not pass condensate, check that the maximum operating pressure marked on the trap is not lower than the actual pressure to which it is subjected. If it is, the valve cannot open. Install a valve and seat assembly with the correct pressure rating. Make sure that this has sufficient capacity to handle the maximum load.

Also check the internals and ensure that the air vent hole in the bucket is not obstructed, because this could cause air-binding.

4.2.2 Liquid Expansion Thermostatic Trap

If a liquid expansion thermostatic trap blows steam, check for dirt or wear on the valve and seat. If wear has occurred, change the complete set of internals. Remember that this type of trap is not self-adjusting to changes in pressure. If it has been set too close at a high pressure, it may not close off at a lower pressure. Try adjusting the trap to a lower setting, making sure that it does not water-log excessively. If it does not appear to react to temperature, install a complete, new set of internals.

4.2.3 Balanced Pressure Thermostatic Trap

If a balanced pressure thermostatic trap blows steam, isolate the trap and allow it to cool before inspecting it for dirt. If the seat is wire-drawn, replace all the internals, including the thermostatic element. The original parts have probably been strained by the continuous steam blow.

If the valve and seat seem to be in good order, check the element. You should not be able to compress it when it is cool. Any flabbiness of the element indicates failure. Flattening of the convolutions indicates water hammer damage. If the water hammer cannot be eliminated at its source, a stronger type of trap must be used.

If the trap will not pass condensate, the element may be overextended due to excessive internal pressure, making it impossible for the valve to lift off its seat. An overexpanded element could be caused by superheat or by someone opening the trap while it is still very hot and before the vapor inside has had time to condense.

4.2.4 Bimetallic Thermostatic Trap

If a bimetallic thermostatic trap blows steam, check for dirt and wear on the valve. A bimetallic trap has limited power to close because of its method of operation, and the valve may be held off its seat by an accumulation of soft deposit.

This type of trap is usually supplied preset. Check that any locking device on the manual adjustment is still secure. If it is not secure, see if the trap responds to adjustment. If cleaning has no effect, install a complete, new set of internals.

The valve on a bimetallic trap is on the downstream side of the valve orifice, so these traps tend to fail in the open position. Failure to pass cold condensate indicates either gross maladjustment or complete blockage of the valve orifice or built-in strainer.

4.2.5 Float Thermostatic Trap

If a float thermostatic trap blows steam, check the trap for dirt at the main valve and the air vent valve. If the trap has a steam-lock release, check that it is not opened too far. It should be open no more than one quarter of a full turn.

Also make sure that the valve mechanism has not been knocked out of line either by rough handling or water hammer, preventing the valve from seating. Check that the ball float is free to fall without fouling the casing, which would cause the mechanism to hang up.

Test the air vent in the same way as the element of a balanced pressure trap is tested. If the internals of a float trap must be replaced, install a complete set as supplied by the manufacturer.

If the trap will not pass condensate, check that the maximum operating pressure marked on the trap is not lower than the actual pressure to which the trap is subjected. If it is, the valve cannot open. Install a valve and seat assembly with the correct pressure rating. Make sure that this has sufficient capacity to handle the maximum load.

If the ball float is leaking or damaged, it is probably caused by water hammer. Also ensure that the air vent or steam-lock release is working correctly.

4.2.6 Thermodynamic Disc Trap

If a thermodynamic disc trap blows steam, the trap will probably give a continuous series of abrupt discharges. Check the trap and strainer for dirt and wipe the disc and seat clean. If this does not correct the problem, it is possible that the seat faces and disc have become worn. The extent of this wear is evident by the amount of shiny surface that replaces the normal crosshatching of machining.

If records show that thermodynamic traps on one particular installation suffer repeatedly from rapid wear, the trap may be oversized, associated pipework may be undersized, or there may be excessive back pressure.

If the trap will not pass condensate, the discharge orifices may be plugged shut with dirt. This is most likely caused by air-binding, particularly if it occurs regularly during start-up. Check the air-venting arrangements of the steam-using equipment. In extreme cases, it may be necessary to install an air vent parallel with the trap or to use a float trap with a built-in air vent instead of a thermodynamic trap.

Review Questions

1. Mechanical steam traps respond to the difference in density between steam and _____.
   a. air
   b. condensate
   c. gases
   d. debris

2. A thermostatic steam trap responds to _____ changes in steam lines.
   a. pressure
   b. density
   c. temperature
   d. flow

3. The float thermostatic trap uses a float to open the main valve and _____ to release cooler gases.
   a. a diaphragm
   b. thermostatic bellows
   c. bimetallic strips
   d. a pressure tap

4. The most widely used thermodynamic steam trap is the _____ type.
   a. bimetallic
   b. disc
   c. balanced pressure
   d. inverted bucket

5. In general, steam trap mountings should be _____ than any line in the system.
   a. smaller
   b. larger
   c. lower
   d. higher

6. A strainer must be located _____ of the steam trap.
   a. outside
   b. inside
   c. downstream
   d. upstream

7. The float thermostatic trap in series with a liquid expansion trap works well in _____ applications.
   a. process heating
   b. power-producing
   c. mechanical
   d. pressure

8. Three basic diagnostic methods of reading a steam system are sight, _____, and temperature.
   a. pressure
   b. sound
   c. flow
   d. drainage

9. The sight method uses a _____ that vents the condensate to the atmosphere for visual inspection.
   a. test valve
   b. vent valve
   c. sight glass
   d. condensate tester

10. If a thermodynamic disc trap blows steam, the trap will probably give a continuous series of _____.
    a. knocking sounds
    b. abrupt discharges
    c. pressure increases
    d. warning whistles

Summary

To maintain the integrity and efficiency of steam use, a steam system must deliver dry steam that is free of condensate, air, gases, and particles. A steam trap and strainer system can deliver clean steam to the equipment. Each type of steam trap has features and limitations that define its use in each application. Proper installation and diligent preventive maintenance will ensure the optimal use of the process steam delivery

Notes

Trade Terms Introduced in This Module

Condensate: The liquid byproduct of cooling steam.

Drip leg: A drain for condensate in a steam line placed at a low point in the line and used with a steam trap.

Flash steam: Steam formed when hot condensate is released to lower pressure and re-evaporated.

Saturated steam: Pure steam without water droplets that is at the boiling temperature of water for the given pressure.

Superheated steam: Saturated steam to which heat has been added to raise the working temperature.

Tracer: A steam line piped beside product piping to keep the product warm or prevent it from freezing.

Water hammer: A condition that occurs when hot steam comes in contact with cooled condensate, builds pressure, and pushes the water through the line at high speeds to slam into valves and other equipment with devastating effect.

Resources & Acknowledgments

Additional Resources

This module is intended to be a thorough resource for task training. The following reference works are suggested for further study. These are optional materials for continued education rather than for task training.

Armstrong Steam Conservation Guidelines for Condensate Drainage, Armstrong Steam Specialty Products, Three Rivers, MI 49093, (616) 273-1415.

Design of Fluid Systems, Steam Utilization, Spirax Sarco Inc., P.O. Box 119, Allentown, PA 18105, (610) 797-5830.

Velan Steam Traps, www.velansteamtraps.com

www.yarway.com/literature.asp

Figure Credits

Armstrong International, Inc., 404F01 (line art), 404F02 (bottom left, bottom right), 404F04 (photo)

Nicholson Steam Trap, 404F01 (photo)

Watson McDaniel, 404F02 (bottom center), 404F03 (photo)

Courtesy of UE Systems, Inc., 404F08

Fluke Corporation, 404F09

Weed Instrument Co., 404F10

NCCER CURRICULA — USER UPDATE

NCCER makes every effort to keep its textbooks up-to-date and free of technical errors. We appreciate your help in this process. If you find an error, a typographical mistake, or an inaccuracy in NCCER's curricula, please fill out this form (or a photocopy), or complete the online form at **www.nccer.org/olf**. Be sure to include the exact module ID number, page number, a detailed description, and your recommended correction. Your input will be brought to the attention of the Authoring Team. Thank you for your assistance.

Instructors – If you have an idea for improving this textbook, or have found that additional materials were necessary to teach this module effectively, please let us know so that we may present your suggestions to the Authoring Team.

NCCER Product Development and Revision
13614 Progress Blvd., Alachua, FL 32615

Email: curriculum@nccer.org
Online: www.nccer.org/olf

❏ Trainee Guide   ❏ AIG   ❏ Exam   ❏ PowerPoints   Other _____

Craft / Level: _____   Copyright Date: _____

Module ID Number / Title: _____

Section Number(s): _____

Description: _____

_____

_____

_____

Recommended Correction: _____

_____

_____

_____

Your Name: _____

Address: _____

_____

Email: _____   Phone: _____

Pipefitting Level Four

08405-07

In-Line Specialties

08405-07
In-Line Specialties

Topics to be presented in this module include:

1.0.0	Introduction	5.2
2.0.0	Safety and Potential Hazards	5.2
3.0.0	Types of In-Line Specialty Equipment	5.2
4.0.0	Storing and Handling In-Line Specialties	5.16

Overview

This module looks at the different kinds of special fittings and instruments used in process piping equipment. In system documentation, these specialties are described as specials. This category includes steam traps, desuperheaters, bursting disks, strainers, and other process equipment. Steam traps and desuperheaters are equipment specialized for steam lines; steam traps protect the lines against water hammer, and desuperheaters reduce the temperature of steam. Strainers are used with many types of fluids, to keep solids from clogging pipes, while bursting disks are a form of emergency pressure relief to prevent very high pressure surges from damaging equipment. The reasons for using a particular piece of equipment are mentioned, as well as particular issues relevant to installation and maintenance.

Objectives

When you have completed this module, you will be able to do the following:

1. Identify the potential hazards associated with in-line specialties.
2. Identify in-line specialties.
3. Explain how to store and handle in-line specialties.

Trade Terms

Conductivity
Differential pressure

Required Trainee Materials

1. Pencil and paper
2. Appropriate personal protective equipment

Prerequisites

Before you begin this module, it is recommended that you successfully complete *Core Curriculum*; *Pipefitting Level One*; *Pipefitting Level Two*; *Pipefitting Level Three*; and *Pipefitting Level Four*, Modules 08401-07 through 08404-07.

This course map shows all of the modules in the fourth level of the Pipefitting curriculum. The suggested training order begins at the bottom and proceeds up. Skill levels increase as you advance on the course map. The local Training Program Sponsor may adjust the training order.

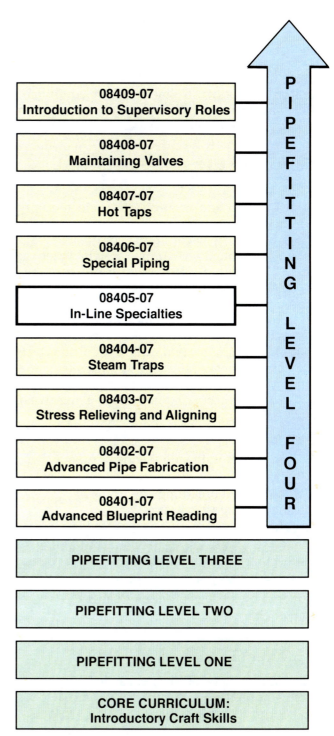

1.0.0 ♦ INTRODUCTION

In-line specialty equipment are devices installed in a piping system for the purpose of metering, measuring, relieving, controlling, filtering, or monitoring product. In-line specialty equipment is delicate in nature, and care must be taken when it is handled to prevent damage.

2.0.0 ♦ SAFETY AND POTENTIAL HAZARDS

The installation and removal of in-line specialty equipment located in process piping systems present a critical safety hazard because you are breaking into a potentially hazardous system. Potential hazards associated with opening process lines include the possibility of being splashed with or exposed to corrosive, toxic, or hot substances, possibly at high pressures. Extreme care must be taken when cutting into or welding on gas lines that contain highly flammable and explosive substances.

Be sure to check the permit requirements and company regulations to see which of the following personal protective equipment you must wear when performing line entries:

- Full-face shield
- Safety glasses with side shields or mono-goggles
- Self-contained breathing apparatus or in-line-supplied respirator
- A complete chemical-resistant suit
- Chemical gloves with cuffs long enough to fit up under the sleeves of the suit
- Rubber boots high enough for the tops to be covered by the legs of the suit pants
- Hard hat

Follow these safety precautions when performing any type of line entry:

- Lock out and tag all electrical switches and disconnects on the equipment as required.
- Before the line is opened, clear and barricade an area large enough to ensure the safety of nearby personnel.
- Provide a container to catch any drained material from the pipeline.
- When a line containing flammable liquids is to be opened, ensure that a dry chemical fire extinguisher and a connected water hose are brought to the area and made ready for use.
- If the equipment is activated by mechanical or control devices, these devices are to be manually operated to be sure that the equipment is indeed taken out of service.
- If the line to be opened contains flammable or hazardous vapors, the line must be purged with nitrogen, water, steam, or air for a long enough period of time (as specified by plant procedures) to ensure that all toxic and flammable vapors have been removed.

3.0.0 ♦ TYPES OF IN-LINE SPECIALTY EQUIPMENT

The following are some common types of in-line specialty equipment:

- Snubbers
- Ball joints
- Bleed rings
- Drip legs
- Expansion joints
- Filters
- Flowmeters
- Level measurement devices
- Flow pressure switches
- Rupture discs
- Thermowells
- Desuperheaters

3.1.0 Snubbers

Snubbers (*Figure 1*) are special fittings that are installed between piping or vessels and pressure gauges. The purpose of snubbers is to protect gauges from pressure surges and shocks in the system. Snubbers smooth and dampen the pressure impulses, thereby prolonging the life and accuracy of the pressure gauges. Pressure dampening or snubbing is accomplished through the use of a porous, stainless steel element assembled into the body of the snubber. This element absorbs and dampens the pressure surges moving through the snubber. Snubbers are widely used with pressure gauges during hydrostatic testing. Without a snubber, the positive displacement action of the test pump causes the gauge needle to move erratically and the gauge to give false readings.

3.2.0 Ball Joints

Ball joints (*Figure 2*) are flanged or welded pipe connections that provide pivot points to allow for expansion and contraction of a piping system. Ball joints solve many movement, alignment, expansion, contraction, vibration, and shock problems in a piping system. Ball joints provide movement in all planes from a single joint. They are designed to swivel or rotate 360 degrees, with 30 to 40 degrees of flexing or angular movement.

3.3.0 Bleed Rings

Bleed rings provide a leakproof means of stopping flow in pipelines. They are used in systems when a change in the process material flowing through the line is made and cross-contamination must be avoided. Bleed rings are installed between the block valves in a pipeline and have a valved bleed connection attached. Before the process flow is changed, the valved bleed connection drains the area between the block valves. *Figure 3* shows a typical pipe arrangement for a bleed ring.

3.4.0 Drip Legs

Drip legs collect condensate and sediments from steam systems. They are made from pipe, fittings, and valves and are installed between the steam header and the equipment at a low point in the steam line. They can be connected directly to the steam header or in the steam line and should be installed at every elevation change in the steam line. At the base of the drip leg is a drip leg drain valve that discharges sediment that collects in the drip leg. A pipeline and an isolation valve are normally attached to the side of a drip leg that is routed to a steam trap. Steam traps that serve a drip leg are installed at a lower level than the drip leg. *Figure 4* shows typical drip leg arrangements.

Figure 2 ◆ Ball joint.

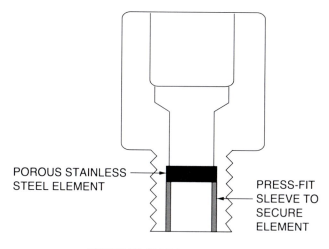

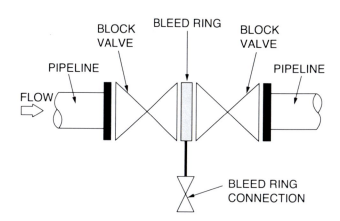

Figure 1 ◆ Snubber.

Figure 3 ◆ Typical pipe arrangement for bleed ring.

MODULE 08405-07 ◆ IN-LINE SPECIALTIES 5.3

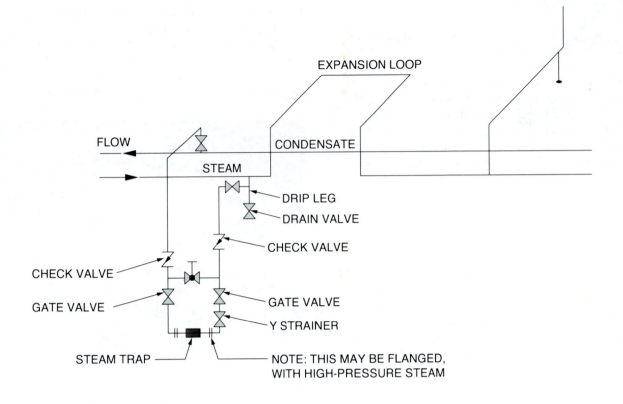
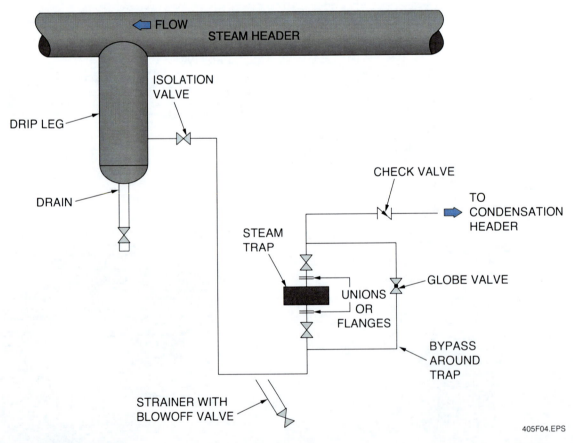

Figure 4 ◆ Drip leg arrangements.

General guidelines for proper drip leg installation are as follows:

- Install the drip leg to the steam header line
- Install a valve at the base of the drip leg to allow condensate to be drained.
- Install a shutoff valve upstream from the drip leg to cut off discharge of condensate from equipment and to allow maintenance work to be performed.
- Install an isolation valve close to the drip leg exit flow to allow maintenance work to be performed without shutting down the supply line.

Steam traps and strainers are located downstream from the drip leg.

3.4.1 Steam Traps

There are three general types of steam traps: mechanical, thermostatic, and thermodynamic (see *Figure 5*). Among the many variations of these are also the various materials used, such as cast iron, stainless steel, forged steel, and cast steel. Steam traps are fitted to the pipe with screwed, flanged, or socket weld fittings.

It is important to have the correct trap for the application. A steam trap of the right size and type will operate reliably, prevent steam loss, and maximize product output. The wrong trap will wear out prematurely and fail to drain the condensate.

All types of steam traps automatically open an orifice, drain the condensate, then close before steam is lost. With the exception of the inverted bucket-type, no steam trap releases any steam when new or in good condition.

General guidelines for proper steam trap installation are as follows:

- Provide a separate trap for each piece of equipment or apparatus. Short-circuiting may occur if more than one piece of equipment is connected to a single trap.
- Tap the steam supply off the top of the steam header to obtain dry steam and avoid steam line condensate.
- Install a supply valve close to the steam header to allow maintenance or revisions to be performed without shutting down the steam header.
- Install a steam supply valve close to the equipment entrance to allow equipment maintenance work to be performed without shutting down the supply line.
- Connect the condensate discharge line to the lowest point in the equipment to avoid water pockets and water hammer.
- Install a shutoff valve upstream of the condensate discharge piping to cut off discharge of condensate from equipment and allow service work to be performed.
- Install a strainer and strainer flush valve ahead of the trap to keep rust, dirt, and scale out of working parts and to allow blow-down removal of foreign material from the strainer basket.
- Provide unions on both sides of the trap for its removal or replacement.
- Install a test valve downstream of the trap to allow observation of discharge when testing.
- Install a check valve downstream of the trap to prevent condensate flow-back during shutdown or in the event of unusual conditions.
- Install a downstream shutoff valve to cut off equipment condensate piping from the main condensate system for maintenance or service work.
- If a bypass around the trap is installed, trap replacement can occur without loss of flow.

3.4.2 Strainers

Strainers (*Figure 6*) stop scale and corrosion in a steam system from entering the trap, thus preventing clogs and damage to the steam trap. In all systems, a strainer should be installed before the steam trap. Strainers are available in many styles and are sometimes incorporated into the steam trap. Strainers must be cleaned and inspected on a regular basis.

Temporary or start-up strainers, also known as witch's hats and top hats (*Figure 7*), are installed during system startup between the first set of flanges upstream from the pump suction to catch any type of construction debris in the system. These strainers can be removed after startup or left in the system permanently, depending on the plant conditions. In the case of conical strainers, whether cone or basket, the strainer should be installed with the smaller end pointed into flow, so that the cone will not fill up too quickly.

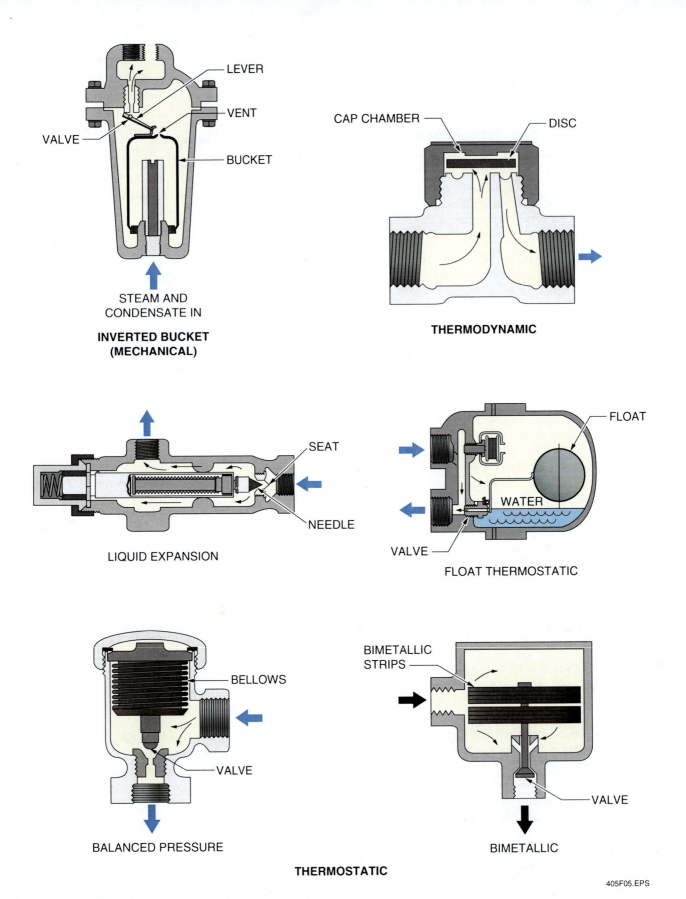

Figure 5 ♦ Types of steam traps.

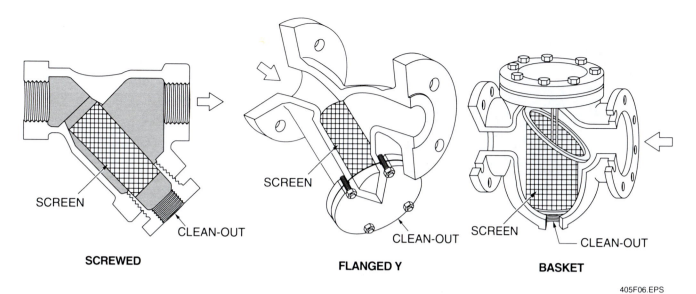

Figure 6 ◆ Strainers.

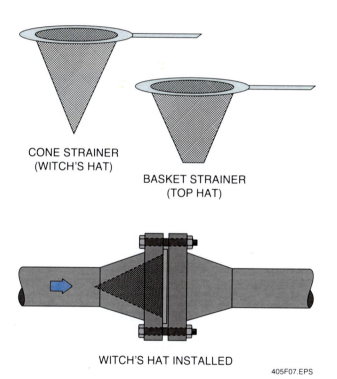

Figure 7 ◆ Temporary strainers.

3.5.0 Expansion Joints

Expansion joints (*Figure 8*) are special pipe fittings designed to control the linear expansion and contraction of the pipe. The differences in the temperature of the fluid carried and the surrounding air make it necessary to fit pipelines with expansion joints to compensate for linear expansion. Charts showing the exact linear expansion for each degree of heat may not always be available. However, the following rule of thumb can be used to calculate the expansion for any length of steel pipe: ¾ inch per 100 feet per 100°.

3.6.0 Filters

Filters collect suspended solids in a system and are located upstream from various devices that benefit from having particles removed from the system. Pressure elements are located upstream and downstream from the filter. These elements take fluid pressure readings at the entrance and discharge of the filter and compare the readings, resulting in a **differential pressure** reading.

As filters age and collect more particles, the differential pressure increases. When the differential pressure increases, it may be necessary to replace the filter. On critical filter systems, a range of differential pressure is established for the selected system, and the range is programmed into the differential pressure monitor. The differential pressure monitor indicates the low end of the range when the filter is clean and indicates the high end of the pressure range when the filter becomes saturated or dirty. When installing filters in a filter housing, make sure you install the correct filter with the proper temperature rating into the filter housing. *Figure 9* shows a filter diagram.

3.7.0 Flowmeters

Flowmeters measure the amount of fluid being transferred in a piping system. There are various types of flowmeters used in piping systems. Methods for measuring flow fall into two categories: invasive and noninvasive. Invasive measurement devices project into a pipe or tube

(A) RUBBER EXPANSION JOINT

(B) FLEXIBLE EXPANSION JOINT

405F08.EPS

Figure 8 ◆ Expansion joints.

that carries material. Noninvasive measurement devices do not enter the pipe or tube that carries material. Because they are outside, noninvasive measurement devices must measure by inference, detecting outside the pipe some property of the material inside the pipe or producing a signal to which the material inside the pipe can react. The following are the four basic types of flowmeters:

- Orifice plates
- Rotameters
- Venturi tubes
- Pitot tubes

3.7.1 Orifice Plates

The orifice plate (*Figure 10*) is the simplest and most commonly used invasive device for flow measurement. An orifice plate is a metal plate installed between two flanges and has a hole smaller than the inside diameter of the pipes in which it is installed.

Orifice plates are used between special flanges that have orifice taps. These tapped holes, located in the flange rims, are used as connection ports for tubing and pressure gauges. Orifice plates are used to determine flow rates by taking pressure readings above and below the plate. The pressure will vary dependent on the rate of flow through the orifice.

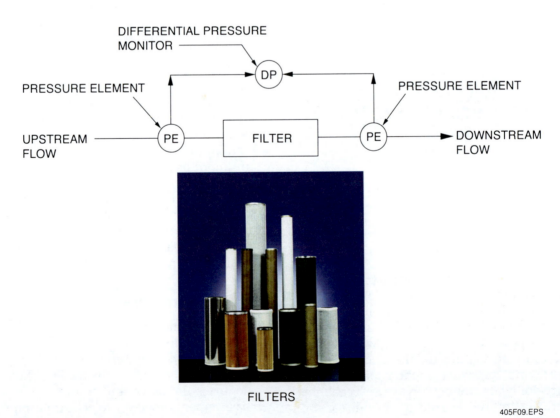

Figure 9 ◆ Filter diagram.

5.8 PIPEFITTING ◆ LEVEL FOUR

All orifice plates have a specific orientation to direction of flow. Also, because of the pressure differential and possible flow characteristic effects, they are to be installed, unless otherwise specified, with a straight horizontal run of pipe that is ten feet long upstream of the plate and five feet long below the plate.

3.7.2 Rotameters

A rotameter (*Figure 11*) is a type of flowmeter that works by measuring and reporting the turning speed of a propeller or turbine placed in the flow of fluid. One type of rotameter consists of a transparent tube with a tapered and calibrated bore, arranged vertically, wide end up, supported in a casing or framework with end connections. The instrument is connected so that the flow enters at the lower end and leaves at the top. A ball or spinner rides on the moving gas inside the tapered tube. As the flow rate increases, the ball or spinner is lifted higher to indicate the flow rate through the rotameter. Isolation and bypass valves are used in conjunction with a rotameter. Another type of rotameter is inserted directly into the pipe from the side with a propeller or turbine protruding into the flow.

3.7.3 Venturi Tubes

A venturi tube is a tube whose diameter is narrower in the center to restrict the flow through the pipe (see *Figure 12*). The venturi tube also recovers pressure drop downstream from the restriction and therefore is considered highly efficient in terms of conserving energy. At the front end of the venturi tube is a suction tap. The sloped area leading up to the restriction in a venturi tube is called the approach area; the restriction is called the throat, and the tube following the throat is called the recovery area. The throat tap is located at the throat.

The throat usually contains one of the pressure-measuring pickup points, and because of the long slope behind the restriction, most of the pressure drop is recovered. The pressure drop reading can be viewed at the recovery tap.

Venturi tubes are used mainly on special applications because special piping considerations make them expensive.

3.7.4 Pitot Tubes

The pitot tube (*Figure 13*) is a flow-measurement device that is used to monitor and meter fluid and steam flow in relatively large pipes. It can be used for a variety of difficult fluid services, including high-temperature, high-pressure, and corrosive materials.

Two types of pitot tubes are the screw type and the insertion type. The insertion type is generally mounted into the pipeline through a valve and then pressure-fitted to the pipeline. This allows the pitot tube to be removed, inspected, and repaired without interrupting the process flow.

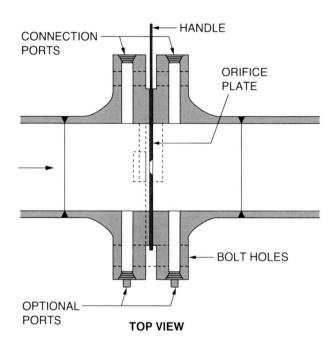

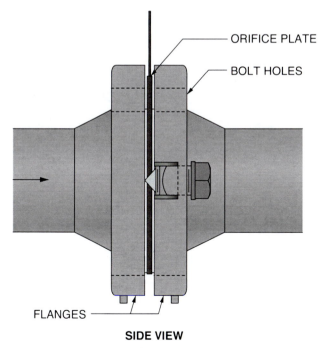

Figure 10 ◆ Orifice plate.

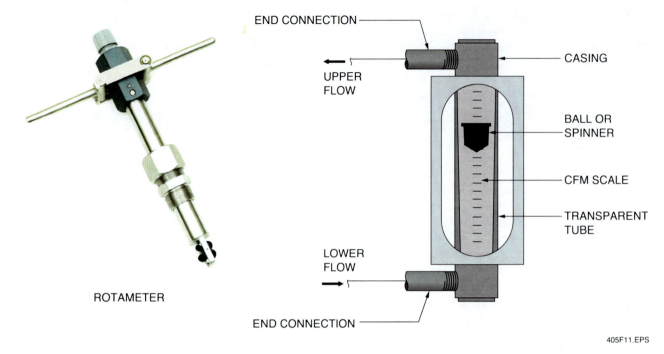

Figure 11 ♦ Rotameters.

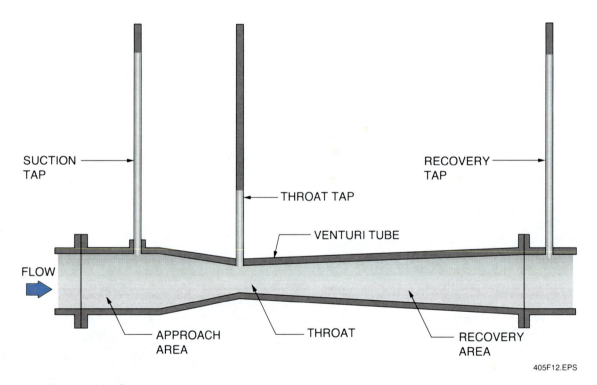

Figure 12 ♦ Venturi tube flow.

3.8.0 Level Measurement Devices

Level measurement devices are sensing instruments used in determining the level of material in a vessel. The fluid level of the contents in vessels is critical in many processes. Two types of level measurement devices are **conductivity** probes and sight glasses.

3.8.1 Conductivity Probes

Conductivity probes (*Figure 14*) measure the conductive characteristics of various processes throughout a job site. Typical applications include rinse tanks, water and waste treatment, condensate monitoring, and desalination units. The conductivity probe is inserted into the pipeline and

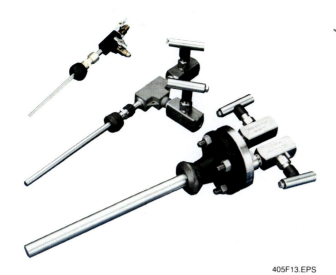

Figure 13 ◆ Pitot tubes.

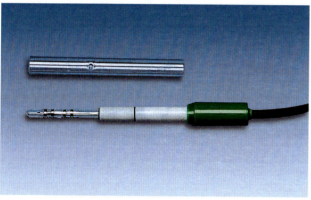

Figure 14 ◆ Conductivity probe.

provides a measurement signal to a conductivity transmitter. The electrode of the conductivity probe must be wetted at all times by the service within the pipeline to provide reliable and accurate measurement signals.

3.8.2 Sight Glasses

Sight glasses give a visual indication of the liquid level in a vessel or tank for both routine observations and level controller adjustments. Sight glasses are usually attached to a pipe that is connected to the vessel. Sight glasses come in a variety of lengths and can be installed in staggered positions, so that when the liquid level exceeds the range of the glass, it is immediately indicated in the next glass. The combination of glasses must cover the complete float range required by the process. *Figure 15* shows typical sight glass arrangements.

3.9.0 Flow Pressure Switches

Flow pressure switches control liquid transfer from one storage area to another. The transfer of liquid often takes a long time; therefore, it is advantageous to provide automatic control of the operation. Once the transfer process has begun, the flow pressure switch takes over and automatically controls the operation until it is finished. Flow pressure switches are installed in the pipeline between the supply tank and the transfer pump. After the transfer pump has been started, the flow pressure switch keeps the pump operating until the transfer of liquid is complete.

Pressure differential transmitters (*Figure 16*) measure flow or level pressures. The measured span of pressure may vary from 0 to 1,000 inches of pressure, depending on the nature of the application. The electronics within the pressure differential transmitter are the same for all pressure differential transmitter applications; however, the physical dimensions of the sensing element vary in proportion to the pressure range of the transmitter.

The sensing element measures the pressure of the flow or level and provides an input signal to the transmitter. The transmitter contains an electronic assembly that receives the input signal and converts it to a proportional output signal.

3.10.0 Rupture Discs

A rupture disc is a safety device that is designed to burst at a certain operating pressure to release gas or liquid from a system. Rupture discs are usually a replaceable metal disc that is held between two flanges. Rupture discs isolate and protect valves from corrosive process fluids. They also serve as supplemental pressure-relief valves.

 WARNING!
Before installing a rupture disk, make sure you understand the direction of flow.

Normally, the installation of a rupture disc is a hold point, and the installation of rupture discs must be witnessed by a quality control inspector or plant representative. Rupture discs are always installed after all tests of the system have been completed. *Figure 17* shows a typical rupture disc setup.

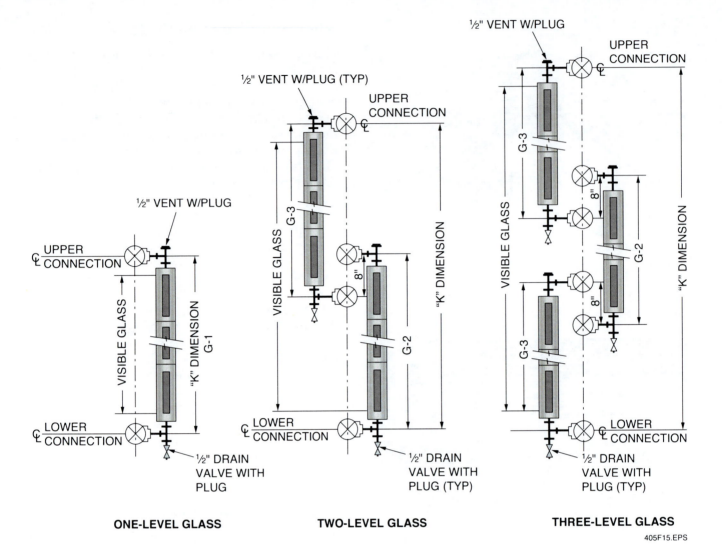

Figure 15 ◆ Sight glass arrangements.

Figure 16 ◆ Pressure differential transmitter.

5.12　PIPEFITTING ◆ LEVEL FOUR

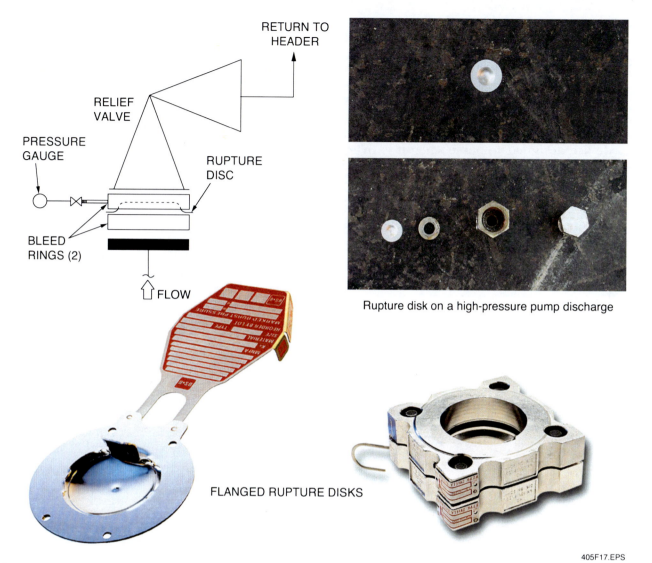

Figure 17 ◆ Rupture disc setup.

There are several types of rupture discs, including the following:

- Conventional
- Scored tension-loaded
- Composite
- Reverse-acting
- Graphite

3.10.1 Conventional Rupture Discs

The conventional rupture disc is a prebulged solid metal disc designed to burst when it is overpressurized on the concave side. The conventional disc of flat seat or angular design is used when operating conditions are 70 percent or less of the rated burst pressure of the disc and when limited pressure cycling and temperature variations are present. For vacuum or back pressure conditions, the conventional disc should be furnished with a support to prevent flexing or implosion.

3.10.2 Scored Tension-Loaded Rupture Discs

The scored tension-loaded rupture disc allows a close ratio of system-operating pressure to burst pressure. It is thicker than the conventional disc and opens along scored lines. The scored tension-loaded disc does not fragment and withstands full vacuum without the addition of a separate vacuum support.

3.10.3 Composite Rupture Discs

The composite disk usually consists of a metal disk that is designed to fail at a certain pressure, with a seal made of some other leak-preventing material on the pressure side. It is designed to burst when it is overpressurized on the concave side. The flat disc is designed to burst when it is overpressurized on the side designated by the manufacturer. Composite rupture discs are available in burst pressures lower than the conventional rupture disc and may offer a longer service life as a result of the corrosion-resistant properties of the seal material selected.

3.10.4 Reverse-Acting Rupture Discs

A reverse-acting rupture disc is a domed solid metal device designed to burst when it is over-pressurized on the convex side. It allows up to 90 percent of the rated burst pressure. Reverse-acting rupture discs do not fragment and do not need vacuum support. They provide good service under vacuum pressure cycling conditions and temperature fluctuations.

3.10.5 Graphite Rupture Discs

A graphite rupture disc is made from graphite impregnated material with a binder and is designed to burst by bending or shearing. It is resistant to most acids, alkalines, and organic solvents and operates to 70 percent of the rated burst pressure. A support is required for a disc that is rated 15 psig or less and in systems with high back pressures. When a disc ruptures, a metal screen or similar catchment for capturing fragmentation must be provided.

3.11.0 Thermowells

Thermowells (*Figure 18*) are metal housings that hold temperature-measuring devices and protect them from the process environment. Thermowells are placed so that they are in contact with the fluid. The main causes of thermowell failure are mechanical failure, corrosion, erosion, and vibration fatigue. Mechanical failures are the result of breakage or bending due to applied force that exceeds the thermowell's yield strength. Corrosion is caused by chemicals and elevated temperatures. Erosion is caused by particles in the system that strike the thermowell at high speeds. Vibration fatigue is caused by the turbulent wake produced when the fluid flows past the thermowell; when this wake frequency equals or exceeds its natural frequency, the thermowell vibrates to its breaking point.

3.12.0 Desuperheaters

Desuperheaters are used to bring the temperature of superheated steam down closer to saturated steam temperature, so that the pressure can also be lowered. There are several varieties of desuperheaters, some involving large tanks or heat exchanger-type tube assemblies, designed

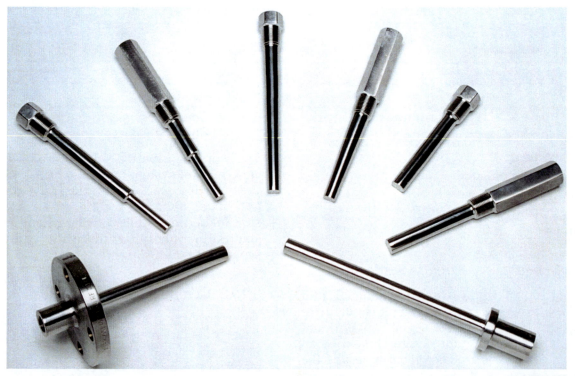

Figure 18 ♦ Thermowells.

to use substantial quantities of cooling water to lower the temperature of superheated steam. Another variety of desuperheaters is the water spray desuperheater, which makes cooling water into very tiny droplets, increasing the surface area, and injects them into the superheated steam. The water uses heat from the steam and evaporates immediately, cooling the steam.

3.12.1 Large Tank Desuperheaters

The large tank variety of desuperheater, also called a direct contact desuperheater, injects the steam into a large bath of water inside a tank, with a steam takeoff leading out of the dry steam space above the water. Some of the water evaporates, removing energy from the superheated steam. The water level is controlled, and makeup water is added to keep the water level constant. These systems allow large quantities of steam to be processed and are relatively simple to operate and maintain. They are bulky, however, and involve a large initial cost.

3.12.2 Heat Exchanger Desuperheaters

The heat exchanger/tube bundle type desuperheater has two shell and tube bundle heat exchangers immersed in cooling water. The steam flows through the tubes, and evaporant from the cooling water escapes as steam from the floating head on one side of the shell. These systems can also process a great deal of steam, and can handle very high temperatures and pressures. Like the water bath type, these are expensive and bulky units, and the large surface area of the tubes can result in scale buildup that decreases the efficiency of the heat transfer.

3.12.1 Water Spray Desuperheaters

Water spray desuperheaters operate within the steam line. There are several varieties, some spraying axially, some injected radially, some with several nozzles, some with variable nozzles. Some are bolted in between two flanges; others are inserted through a hole in the side of the pipe (*Figure 19*). Some are sprayed axially against the flow of steam, increasing the turbulence and time of contact between the steam and water.

One type of water spray desuperheater is a dual-fluid, tube-based nozzle positioned at a 45-degree angle to a steam line (see *Figure 20*).

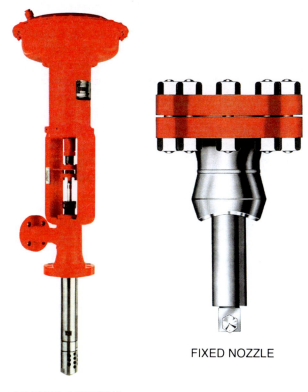

Figure 19 ♦ Insert spray desuperheaters.

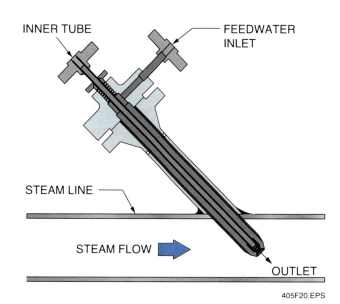

Figure 20 ♦ Desuperheater.

Two entry valves positioned 90 degrees apart are located on the input side. The line on the end of the tube allows steam to enter an inner tube of the desuperheater. The feedwater inlet on the side of the desuperheater shaft allows condensate to enter the other tube of the desuperheater. Both fluids travel down the length of the desuperheater and are directed into a number of flow channels at the nozzle end. The flow channels for the steam are sized to produce a near critical pressure drop. This creates very high steam velocity and maximum energy for spray water atomization.

The atomizing steam disperses the cooling condensate into fine droplets, which are then injected into the superheated steam in the main pipe. This spray cloud, consisting of a very large surface area, is vaporized almost instantaneously by the superheated steam to produce steam at a lower temperature. The amount of condensate introduced into the main steam line determines the final temperature of the steam.

4.0.0 ◆ STORING AND HANDLING IN-LINE SPECIALTIES

The proper functioning of in-line specialty equipment depends not only on the inherent quality of the instrument itself but also on how well the equipment is stored and handled. The following guidelines should be followed when storing and handling in-line specialty equipment:

- Store pitot tubes in reinforced boxes.
- Keep pitot tubes clean.
- Keep ball joints crated until they are installed.
- Handle conductivity probes with extreme care.
- Do not flex, puncture, or contaminate filters.
- Dispose of contaminated filters according to EPA regulations.
- Store flow pressure switches in a dry area and in their original packing until they are ready to be installed.
- Handle flow pressure switches with extreme care.
- Do not dent, scratch, or alter orifice plates.
- Do not strain, flex, or bend rupture discs.
- Store rupture discs in their packing until they are ready to be installed.
- Do not remove any identification tags or plates from any in-line equipment.

Review Questions

1. Snubbers smooth and dampen _____.
   a. heat waves
   b. steam surges
   c. pressure impulses
   d. process flow

2. Ball joints are installed to allow for _____ in a piping system.
   a. expansion and contraction
   b. chemical absorption
   c. pressure relief
   d. opening and closing

3. Drip legs collect condensate and sediments from _____ systems.
   a. hot-water
   b. process water
   c. gas
   d. steam

4. The most commonly used invasive device for measuring flow is a(n) _____.
   a. rotameter
   b. venturi tube
   c. pitot tube
   d. orifice plate

5. Which type of flowmeter uses propellers and turbines inserted in the flow of fluid?
   a. Rotameter
   b. Venturi tube
   c. Pitot tube
   d. Orifice plate

6. Which of the following is a level measurement device?
   a. Pitot tube
   b. Sight glass
   c. Thermowell
   d. Flowmeter

7. The conventional rupture disc is designed to burst when overpressurized on the _____.
   a. convex side
   b. concave side
   c. top face
   d. bottom face

8. A desuperheater is designed to _____ of steam.
   a. increase the pressure
   b. decrease the acidity
   c. increase the temperature
   d. decrease the temperature

9. Until they are ready to be installed, flow pressure switches should be stored in a(n) _____.
   a. saline solution
   b. dry area
   c. damp area
   d. open area

10. Pitot tubes should be stored in _____.
    a. reinforced boxes
    b. dry areas
    c. crates
    d. packing from other specialties

Summary

In-line specialties are used in many of the process systems that pipefitters work with. These devices serve critical functions in the system and must be installed correctly to operate properly. Many of the items discussed in this module are instrumentation devices that you may or may not be responsible for. The purpose of this module is to make you aware of their purpose so you can recognize these devices in the field.

Notes

Trade Terms Introduced in This Module

Conductivity: A measure of the ability of a material to transmit electron flow.

Differential pressure: The measurement of one pressure with respect to another pressure, or the difference between two pressures.

Resources & Acknowledgments

Additional Resources

This module is intended to present thorough resources for task training. The following reference works are suggested for further study. These are optional materials for continued education rather than for task training.

www.yarway.com/literature.asp

www.spiraxsarco.com/resources/steam-engineering-tutorials.asp

Figure Credits

Ashcroft, Inc., 405F01 (photo)

EBAA Iron, Inc., 405F02, 405F08 (bottom)

The Mack Iron Works Company, Sandusky, Ohio, 405F03 (photo)

Flexicraft Industries, 405F08 (top)

Filter-Fab Corporation, 405F09 (photo)

Turbo-Flo meter for water flow measurement in 1.5–72 pipe sizes, **ERDCO Engineering Corp**, Evanston, IL, 405F11 (photo)

Meriam Process Technologies, 405F13

HANNA® Instruments USA, 405F14

Photo courtesy of **Emerson Process Management and North American Manufacturing**, 405F16

Ancon Marine, 405F17 (top right, bottom left)

BS&B Safety Systems, 405F17 (bottom right)

TEMPCO Electric Heater Corporation, 405F18

Courtesy of **SPX Cooling Technologies, Inc.**, 405F19

NCCER CURRICULA — USER UPDATE

NCCER makes every effort to keep its textbooks up-to-date and free of technical errors. We appreciate your help in this process. If you find an error, a typographical mistake, or an inaccuracy in NCCER's curricula, please fill out this form (or a photocopy), or complete the online form at **www.nccer.org/olf**. Be sure to include the exact module ID number, page number, a detailed description, and your recommended correction. Your input will be brought to the attention of the Authoring Team. Thank you for your assistance.

Instructors – If you have an idea for improving this textbook, or have found that additional materials were necessary to teach this module effectively, please let us know so that we may present your suggestions to the Authoring Team.

NCCER Product Development and Revision
13614 Progress Blvd., Alachua, FL 32615

Email: curriculum@nccer.org
Online: www.nccer.org/olf

❑ Trainee Guide ❑ AIG ❑ Exam ❑ PowerPoints Other _____

Craft / Level: _____ Copyright Date: _____

Module ID Number / Title: _____

Section Number(s): _____

Description: _____

_____

_____

_____

_____

Recommended Correction: _____

_____

_____

_____

_____

Your Name: _____

Address: _____

_____

Email: _____ Phone: _____

Pipefitting Level Four

08406-07

Special Piping

08406-07
Special Piping

Topics to be presented in this module include:

1.0.0	Introduction	6.2
2.0.0	Installing Flared and Compression Joints Using Copper Tubing	6.2
3.0.0	Soldering and Brazing Copper Tubing and Fittings	6.7
4.0.0	Bending Pipe	6.14
5.0.0	Glass-Lined Piping	6.21
6.0.0	Hydraulic Fitted Compression Joints	6.24
7.0.0	Grooved Piping Systems	6.29

Overview

This module introduces the special skills and technologies associated with small pipe and tubing. The connections with copper, stainless steel, aluminum, and brass pipe and fittings are covered. Instructions are provided on brazing, soldering, compression fittings, and flared fittings. The procedure is described for applying Lokring® and other technologies used for assembling air and water lines and instrument connections. These systems are used for field routing of small pipe and instrumentation lines.

Objectives

When you have completed this module, you will be able to do the following:

1. Install flared and compression joints, using copper tubing.
2. Solder and braze joints, using copper tubing.
3. Bend pipe to a specified radius.
4. Install glass-lined pipe.
5. Explain how to install hydraulic fitted compression joints.
6. Install grooved pipe couplings.

Trade Terms

Acetylene
Compression coupling
Ferrule
Flaring
Nonferrous metal
Oxidation
Reamer
Solder
Soldering flux
Wetting

Required Trainee Materials

1. Pencil and paper
2. Appropriate personal protective equipment

Prerequisites

Before you begin this module, it is recommended that you successfully complete *Core Curriculum*; *Pipefitting Level One*; *Pipefitting Level Two*; *Pipefitting Level Three*; and *Pipefitting Level Four*, Modules 08401-07 through 08405-07.

This course map shows all of the modules in the fourth level of the Pipefitting curriculum. The suggested training order begins at the bottom and proceeds up. Skill levels increase as you advance on the course map. The local Training Program Sponsor may adjust the training order.

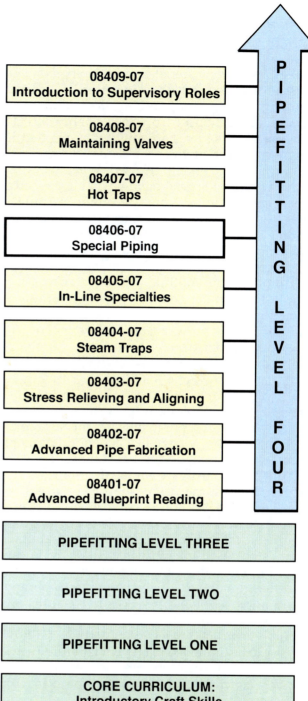

1.0.0 ♦ INTRODUCTION

The term *special piping* refers to a complete network of special pipes, fittings, valves, and other components that are designed to work together to convey a specific material. Special piping systems must be designed and fabricated according to local building codes, plans, specifications, and component installation instructions. Where conflicts between these sources are found, the one that is the most conservative and provides the greatest margin of safety should be used. When working with any type of special piping, you must receive specialized training from the piping manufacturer prior to installing or repairing the pipe. This module explains different types of special piping and their applications, which include assembling, bending, and installing special piping and fittings.

Although this module primarily covers the simplest and most common form of compression fittings, it is important to note that a vast array of fittings are available for every conceivable application. Copper tube fittings are certainly common, but compression fittings are used with plastic, steel, and stainless steel tubing as well, and are applied across many industries. Some utilize multiple ferrules, which are forced by the compression nut to bite into the surface of the tubing with a wedge-like action. Equipment manufactured outside the US is quite likely to be built with hardware under the metric standard, including the use of metric compression fitting systems. Although compression fittings often appear to be the same, it is important that compression hardware from different manufacturers is not used interchangeably. Doing so can result in joint failure and possible injury due to the differing standards among manufacturers and nearly imperceptible differences in construction. Many compression fittings also have very specific assembly instructions regarding torque and how they are to be tightened to prevent failure. Always ensure that compression fitting components are properly matched, and that you are familiar with the assembly instructions associated with the specific product being installed or manipulated. Some compression fittings can be removed and retightened many times, while others cannot.

2.0.0 ♦ INSTALLING FLARED AND COMPRESSION JOINTS USING COPPER TUBING

Copper tubing can be connected by flared or compression joints. Flared joints typically are used when the connection needs to be disassembled and reassembled often, or when the tubing content cannot be removed completely for repairs. They are also used when soldering cannot be used to join the tubing because of the presence of hazardous liquid or fuel within the tube, or when the tubing is run underground. Flared joints are used underground because solder is destroyed by chemicals in the ground.

Compression fittings are used for quick-finish connections that require limited equipment and finish work. Two types of quick-finish connections are made on ice makers and humidifiers that need to be connected to the structure's water supply.

Flared and compression joints are generally used with soft copper tubing, which differs from hard-drawn copper tubing. While hard-drawn tubing is made by drawing the copper through a series of dies to reduce its diameter, soft tubing is heated and slowly cooled, leaving it less brittle, more pliable, and less likely to tear. These characteristics make soft copper tubing an excellent choice for flared or compression joint applications.

Soft copper tubing is offered in three different types, designated by letters: Type K, Type L, and Type ACR. They differ by weight, or wall thickness, and are color-coded for convenience and instant identification. Type ACR and Type L have the same basic characteristics and wall thickness, but Type ACR (for air conditioning and refrigeration use) is purged, pre-charged with nitrogen, and sealed at the point of manufacture to prevent copper oxide from forming. When used in refrigeration circuits, internal cleanliness of the tubing is demanded.

The primary advantage of soft tubing is that it is produced in long lengths, which reduces the number of connections, and it can be bent to fit almost all conditions. Heavier tubing has a thicker wall and can withstand greater pressures.

After the tubing is produced, the manufacturer identifies it by printing the company name or trademark and the tubing type on the entire length of coil. Type K has the thickest wall and is color-coded green. This type of tubing is usually used for underground services or high-pressure applications. Type L and Type ACR have a thinner wall and are both color-coded blue. This type of tubing usually is used for aboveground services, but it can also be used underground.

Many sizes of soft copper are available. Both Types K and L are manufactured in coils of 40, 60, and 100 feet. Coils of 100 feet and 60 feet are manufactured in all sizes up to and including 1 inch. The 60-foot coils are also available in the 2-inch size.

Tubing is sized by its outside diameter (OD) to permit all soldered fittings to be used on any type of tubing. *Table 1* lists copper pipe and tubing specifications.

2.1.0 Flared and Compression Methods

Flared joining consists of placing a flare nut on the end of copper tubing and *flaring* the end of the tubing, using a flaring tool. The flared tubing is then placed against the tapered surface of the fitting, and the flare nut is tightened onto the fitting. Long or short flare nuts are available. Long nuts are used more often on refrigeration and fuel oil piping, where vibration could affect the performance. The flare nut forces the flared tubing and the fitting into a leakproof seal. An advantage of this assembly is that it can be easily disassembled if it is necessary to make repairs. *Figure 1* illustrates a typical flare joint, using a long flare nut, while *Figure 2* shows a typical short and long flare nut for comparison.

In some applications, the angle of the flare formed on the tubing may be different. Most situations will require a flare angle of 45 degrees, but 37-degree flare angles exist as well. The Society of Automotive Engineers primarily governs the standards of flare fittings and their construction. In other unique cases, you may also encounter double-flare fittings, made by turning the end of the tubing inward first, then forming it outward, creating a flared opening of double thickness.

Compression fittings are made up of a threaded body, a threaded nut, and a brass or plastic ring or sleeve called a ferrule. The compression ring, or ferrule, is compressed between the nut and the fitting to produce a leakproof seal. This compression method is quick and easy and requires few tools or equipment. Compression joining makes it easy to attach humidifiers, ice makers, and other fixtures to a water system. *Figure 3* shows a compression joint.

2.2.0 Fittings

The same types of fittings are available for flared and compression methods of joining. To select and install fittings, you must understand the sizing, labeling, and types of fittings.

Table 1 Copper Pipe and Tubing Specifications

TYPE K		
Nominal Size (inches)	Actual OD	Wall Thickness
¼	0.375	0.035
⅜	0.5	0.049
½	0.625	0.049
⅝	0.75	0.049
¾	0.875	0.065
1	1.125	0.065
1¼	1.375	0.065
1½	1.625	0.072
2	2.125	0.083
2½	2.625	0.095
3	3.125	0.109
3½	3.625	0.12
4	4.125	0.134
5	5.125	0.16
6	6.125	0.192
TYPE L		
Nominal Size (inches)	Actual OD	Wall Thickness
¼	0.375	0.03
⅜	0.5	0.035
½	0.625	0.04
⅝	0.75	0.042
¾	0.875	0.045
1	1.125	0.05
1¼	1.375	0.055
1½	1.625	0.06
2	2.125	0.07
2½	2.625	0.08
3	3.125	0.09
3½	3.625	0.1
4	4.125	0.114
5	5.125	0.125
6	6.125	0.14

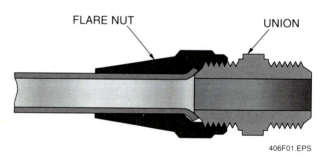

Figure 1 ◆ Flared joint.

Figure 2 ◆ Typical flare nuts.

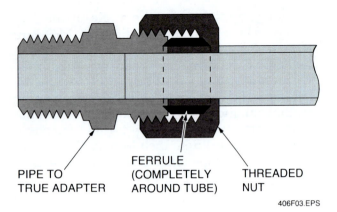

Figure 3 ♦ Compression joint.

2.2.1 Sizing of Fittings

Tube sizes are used to determine the size of fittings. Tube sizes, set by the American Society for Testing and Materials International (ASTM), are nominal. A nominal size is one that is bigger or smaller than the actual measurement. People generally refer to nominal sizes, even though actual size is different.

In refrigeration, tube size is determined by measuring the OD of the pipe. The measured size of the tube is always 1/8 inch larger than the nominal size of the tube. The inside diameter (ID) varies because of wall thickness. As a result, Type K, which is thick-walled tubing, has a smaller ID than Type L tubing.

2.2.2 Labeling of Fittings

Industrial tubing and fittings are labeled by manufacturers, the labels being based on a scheme developed by each manufacturer. The labels usually identify the size, type, and alloy. These systems have not been nationally systematized or standardized. If you work with a particular manufacturer's products for any length of time, you will come to understand their labels.

2.2.3 Types of Fittings

Many types of fittings (*Figure 4*) are manufactured for flared or compression joints. The most common fittings are the following:

- Tee
- 90-degree elbow (ell)
- 45-degree elbow
- Union
- Cross

In addition to the fittings listed above, special fittings are available to connect flared and compression tubing to brazed, soldered, plastic, or threaded pipe. The only type of material that cannot be used with compression fittings is glass-lined pipe.

Flared and compression fittings should be ordered by their OD dimensions, not their ID dimensions. For example, a coupling with a 3/8-inch ID tubing takes a 1/2-inch **compression coupling**.

2.3.0 Installing Flared Fittings

Flaring tools (*Figure 5*) are used to spread the ends of tubing joined by flare nuts and fittings. The flared ends of the tubing, along with the flare nuts, provide a method of coupling tubing without welds, solder, or leaks. There are two types of flaring tools commonly used: the hammer-type and the screw-in type.

Flared fittings are for low pressure piping systems. To make a flare joint, the nut of the fitting must be placed on the tube before the flare is made. The tube is then flared, using a flaring tool. The hammer-type flaring tool comes in different sizes for different tube diameters. Once the correct size is found, insert the tool into the end of the tube and drive it in with firm hammer blows. When the correct flare is made, remove the flaring tool and assemble the joint. The following procedure explains how to use the screw-in type flaring tool, also known as a flaring block, and a yoke and screw flaring tool. Follow these steps to install flared fittings, using a screw-in type flaring tool:

Step 1 Identify the size of the tubing to be flared.

Step 2 Select the flaring tool needed for the size of tubing being worked.

Step 3 Inspect the selected flaring tool for obvious damage or excessive wear on the screw tip or the anvil clamping bars, and for grease, rust, or excessive dirt. If any damage is found, replace the flaring tool. If grease, rust, or excessive dirt is found, clean and oil the tool as necessary.

Step 4 Inspect and clean the end of the tubing to be flared.

CAUTION

The end of the tubing must be square and free of burrs, both inside and outside, or the flared tubing may cause leaks when used. Correct any problem before continuing.

Step 5 Place one of the flare nuts over the end of the tubing and out of the way, with the threaded end facing in the same direction as the tubing end.

Step 6 Loosen the locking nut on the anvil clamping bars.

Step 7 Open the bars wide enough to allow the tubing end to be placed inside one of the flared holes.

Step 8 Move the end of the selected tubing between the open bars and into the flared hole closest to the tubing diameter.

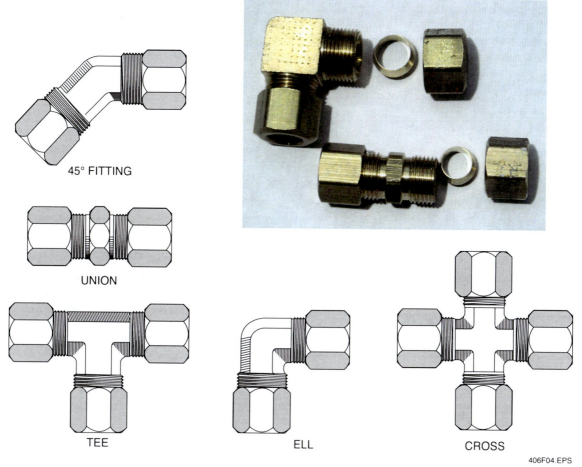

Figure 4 ◆ Fittings.

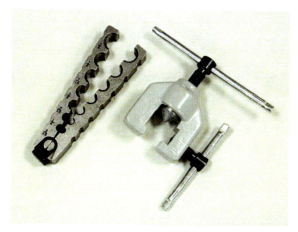

FLARING TOOL

HAMMER-TYPE FLARING TOOL

Figure 5 ◆ Flaring tools.

MODULE 08406-07 ◆ SPECIAL PIPING 6.5

Step 9 Close the bars loosely on the tubing.

Step 10 Position the tubing so that the end extends toward the flaring screw tip above the face of the bars by a distance equal to about half the tubing diameter. Allow for less than half a diameter extension for tubing of ½-inch diameter or more.

Step 11 Tighten the locking nut on the clamping bars while holding the tubing in position. It is important to tighten the bars thoroughly so that the tubing cannot twist or slip while it is being flared.

Step 12 Position the flaring pin screw tip over the end of the tubing. The screw may have to be raised before it is positioned over the tube end. A drop of light oil may be used on the tip to reduce the danger of tube splitting.

Step 13 Tighten the flaring screw down into the end of the tubing to cause the tubing to spread, and stop tightening when the tubing end is flared to an angle of approximately 45 degrees.

CAUTION
Do not overtighten, or the tubing may split as it flares.

Step 14 Unscrew and remove the flaring screw tip.

Step 15 Loosen the locking nut, and open the clamping bars.

Step 16 Remove the flared tubing from the tool.

Step 17 Inspect the tubing flare.

CAUTION
The flared tubing must be evenly spread without any tears or splits. Flared tubing with flaws will result in a leaky joint.

Step 18 Slide the flare nut over the flared end until the flared end seats inside the connector. The flared end should appear well-matched to the inside of the connector.

Step 19 Screw the fitting into the flare nut.

Step 20 Hold the fitting, using a combination wrench, and use another combination wrench to tighten the flare nut onto the fitting.

2.4.0 Installing Compression Fittings

CAUTION
Never mix compression fitting components of different manufacturers. Ensure that all components, including fittings, nuts, and ferrules are compatible and adhere to manufacturer-specific assembly instructions for the type of fittings being used.

Follow these steps to install a compression fitting.

Step 1 Identify the size of the tubing to be fitted.

Step 2 Clean and ream the end of the tubing.

CAUTION
The end of the tubing must be square and free of burrs, both inside and outside, or the compression joint may cause leaks when used. Correct any problem before continuing.

Step 3 Place the fitting over the end of the tubing and out of the way, with the threaded end facing in the same direction as the tubing end.

Step 4 Place a ferrule over the end of the tubing, ensuring that it is properly oriented for the style of fitting being installed.

Step 5 Slide the tubing end into the appropriate fitting. Take care to ensure the tubing end is squarely bottomed out inside the fitting and remains in that position during the remainder of the assembly process.

Step 6 Slide the ferrule into position against the fitting.

Step 7 Screw the threaded compression nut onto the fitting and firmly hand-tighten.

Step 8 Hold the fitting using one combination wrench, while turning the compression nut with a second combination wrench. The nut should be tightened 1¼ turns beyond firmly hand-tight. Do not overtighten the nut, as that will produce a leak. There are go-no-go gauges for checking fittings.

3.0.0 ◆ SOLDERING AND BRAZING COPPER TUBING AND FITTINGS

Soldering and brazing are two other methods used for joining copper tubing and fittings. Soldered joints are used for domestic water lines, sanitary drain lines, hot water heating systems, and other applications in which the temperature does not exceed 250°F. Brazed joints are used where greater strength is required or where temperatures do exceed 250°F. Brazing is used in some steam/hot water heating lines, refrigeration lines, compressed air lines, vacuum lines, oil lines, and some chemical lines that need extra corrosion-resistance in the piping joints. The main difference in the soldering and brazing procedures is the temperature at which each is performed. Most often, soldering is performed at temperatures below 800°F; brazing is performed at temperatures greater than 800°F.

When assembling long runs of tubing, the tubing must be properly supported to eliminate any danger of poor alignment. Both vertical and horizontal support requirements are governed by code. For example, for vertical support of copper tubing systems, both the *Uniform Plumbing Code* and the *BOCA National Plumbing Code* require support at maximum intervals of 10 feet. The *Standard Plumbing Code* requires support at each story for tubing equal to or greater than 1½ inches in diameter. For tubing equal to or less than 1¼ inches in diameter, support must be provided at maximum intervals of 4 feet.

For horizontal support, both the *Standard Plumbing Code* and the *Uniform Plumbing Code* require support at 6-foot intervals for tubing equal to or less than 1½ inches in diameter. For tubing greater than 1½ inches in diameter, horizontal support must be provided at 10-foot intervals. The *BOCA National Plumbing Code* requires horizontal support at 6-foot intervals for tubing equal to or less than 1¼ inches in diameter. For diameters greater than 1½ inches, horizontal support must be provided at 10-foot intervals.

The trend here is obvious—tubing of a smaller diameter requires more frequent support than tubing of a larger diameter due to the structural integrity of the tubing. Proper support of tubing and piping near any change in direction is also critical. It is important to note that these are minimum standards. Local codes and job specifications should always be consulted to ensure compliance. In many situations, more support may be necessary.

It is important to check for proper alignment. Poor alignment affects joint integrity by changing the space and gap between the tubing and fitting walls, and also leaves a poor final appearance. Wires or other temporary hangers can be used to achieve proper alignment during installation.

3.1.0 Soldering Copper Tubing and Fittings

Soldering is the most common method of joining copper tubing and fittings. Soldering involves joining two metal surfaces by using heat and a **nonferrous metal**. A nonferrous metal is a metal that contains no iron and is therefore nonmagnetic. The melting point of the filler metal must be lower than that of the two metals that are being joined.

Soldered joints depend on capillary action to pull the melted solder into the small gap between the fitting and the tubing. Capillary action is the flow of liquid, in this case solder, into a small space between two surfaces. Capillary action is most effective when the space between the tubing and the fitting is between 0.002 and 0.005 inch. *Figure 6* shows capillary action.

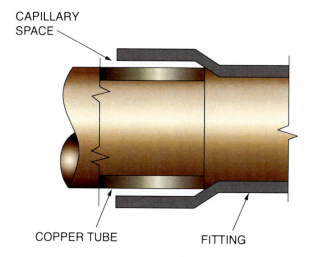

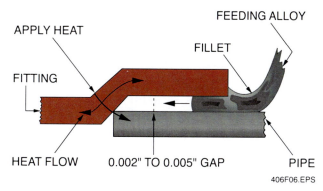

Figure 6 ◆ Capillary action.

To properly solder copper tubing and fittings, you must understand the following materials and procedures:

- Solders and soldering fluxes
- Preparing tubing and fittings for soldering
- Soldering joints

3.1.1 Solders and Soldering Fluxes

Solder is a nonferrous metal or metal alloy with a melting point below 800°F. An alloy is any substance made up of two or more metals. The most common solder used on copper tubing is an alloy made of 95 percent tin and 5 percent antimony. This type of solder is used for potable water connections because it contains no lead. It is usually recommended for applications requiring greater joint strength. The tin-antimony alloy solder melts at 464°F and solidifies rapidly.

The *Federal Safe Drinking Water Act of 1986* mandated the use of lead-free (less than 0.2 percent) solders for drinking water supply piping. Other solders containing antimony and cadmium also should be avoided for drinking water applications. Solders commonly applied in both drinking water and refrigerant piping applications include 95.5 percent tin/4 percent copper/0.5 percent silver and 96 percent tin/4 percent silver. The addition of silver to the solder increases overall strength significantly and, due to its elongation properties, works well where temperature and pressure causes expansion of the tubing. Other choices include those with nickel added for extra strength. Indeed, the proper use and selection of low temperature solders can often result in a superior joint when compared to high-temperature brazing. Excessive heat applied during brazing can further anneal copper, weakening the metal.

Many other solders are available for various applications, including 95 percent tin/5 percent antimony commonly used today on low-pressure refrigeration piping and non-drinking water applications. Fifty percent tin/50 percent lead is still readily available for non-drinking water applications, but 95/5 can withstand roughly 2.5 to 3 times higher working pressures at temperatures up to 250°F. It should be noted that applications over 250°F generally require brazing instead of soldering to ensure joint integrity and strength.

Before selecting solder for a given application, it is essential to understand the proper use and application of the product and ensure that its composition meets the required standards. Technical specification sheets and MSDS reports are readily available for all such products and should be studied carefully when using a new product.

Choosing the proper flux is very important. Soldering flux performs many functions, and the wrong flux can ruin the soldered joint. Flux performs the following functions:

- Flux chemically cleans and protects the surfaces of the tubing and its fitting from oxidation. Oxidation occurs when the oxygen in the air combines with the recently cleaned metal. Oxidation produces tarnish or rust in metal and prevents solder from adhering.
- Flux allows the soldering alloy or filler metal to flow easily into the joint.
- Flux floats out remaining oxides ahead of the molten filler metal.
- Flux promotes wetting of the metals. Wetting is the process that decreases the surface tension so that the molten solder flows evenly throughout the joint.

Fluxes can be classified into three general groups: highly corrosive, less corrosive, and noncorrosive. All fluxes need to be somewhat corrosive. However, the fluxing process must render the flux inert, or lacking any chemical action. If not rendered inert, the flux gradually destroys the soldered joint. As is the case with solder, many different types of flux are available. Solder-flux pastes are also available, containing finely ground particles of a particular solder mixed with a paste flux. The addition of filler metal during the soldering process is still generally required however, dependent upon the application and product.

Although not an absolute requirement, it is suggested that a flux be chosen from the same manufacturer as the chosen solder. Although many solder and flux products are interchangeable, using a flux provided by the solder manufacturer ensures that the two products have been carefully tested together and likely offers the best result.

The best fluxes for joining copper tubing and copper fittings are noncorrosive fluxes. The most common noncorrosive fluxes are compounds of mild concentrations of zinc and ammonium chloride, with petroleum bases.

An oxide film begins forming on copper immediately after it has been mechanically cleaned. Therefore, it is important to apply flux immediately to all recently cleaned copper fittings and tubing. Flux should be applied to the clean metal with a brush or swab, never with fingers. Not only is there a chance of causing infection in a cut, but there is also a chance that flux could be carried to the eyes or mouth. Body contact with the cleaned fittings and tubing adds to unwanted contamination of the metal.

If a can of flux is not closed immediately after use, or if a can is not used for a considerable length of time, the chlorides separate from the petroleum base. Therefore, the flux must be stirred before each use.

CAUTION

Brazing flux and soldering flux are not the same. Do not allow these fluxes to become mixed or interchanged. Carelessness can ruin quality work.

3.1.2 Preparing Tubing and Fittings for Soldering

To prepare tubing and fittings for soldering, the tubing must be measured, cut, and reamed, and the tubing and fittings must be cleaned. It is critical that proper cleaning techniques be used in order to produce a solid, leakproof joint. Follow these steps to prepare the tubing and fittings for soldering.

Step 1 Measure the distance between the faces of the two fittings, using the face-to-face method (*Figure 7*).

Step 2 Determine the cup depth engagement of each of the fittings. The cup depth engagement is the distance that the tubing penetrates the fitting. This measurement can be found by measuring the fitting or by using a manufacturer's makeup chart. *Table 2* shows a manufacturer's makeup chart.

Step 3 Add the cup depth engagement of both fittings to the measurement found in Step 1 to find the length of tubing needed.

Step 4 Cut the copper tube to the correct length using a tubing cutter.

Step 5 Ream the inside and outside of both ends of the copper tube using a **reamer**.

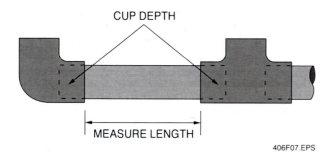

Figure 7 ◆ Face-to-face method.

Table 2 Manufacturer's Makeup Chart

Pipe Size	Depth of Cup	Pipe Size	Depth of Cup
1/4	5/16	2	1 11/32
3/8	3/8	2 1/2	1 15/32
1/2	1/2	3	1 21/32
5/8	5/8	3 1/2	1 29/32
3/4	3/4	4	2 5/32
1	29/32	5	2 21/32
1 1/4	31/32	6	3 3/32
1 1/2	1 3/32	–	–

406T02.EPS

Step 6 Clean the tubing and the fitting using No. 00 steel wool, emery cloth, or a special copper-cleaning tool.

CAUTION

Care must be taken when cleaning copper tubing and fittings to remove all of the abrasions on the copper without removing a large amount of metal. The abrasions can weaken or ruin a copper joint. Do not touch or brush away filings from the tube or fitting with your fingers because your fingers will also contaminate the freshly cleaned metal.

Step 7 Apply flux to the copper tubing and to the inside of the copper fitting socket immediately after cleaning them.

Step 8 Insert the tube into the fitting socket, and push and turn the tube into the socket until the tube touches the inside shoulder of the fitting.

Step 9 Wipe away any excess flux from the joint.

Step 10 Check the tube and fitting for proper alignment before soldering.

3.1.3 Soldering Joints

Because soldering requires relatively low heat, heating equipment that mixes **acetylene**, butane, or propane directly with air is all that is needed. This means that you need only one tank of gas and a torch. Follow these steps to solder a joint.

WARNING!

Always solder in a well-ventilated area because fumes from the flux can irritate your eyes, nose, throat, and lungs.

Step 1 Obtain either an acetylene tank and related equipment or a propane bottle and torch.

Step 2 Set up the equipment according to the manufacturer's instructions.

Step 3 Light the heating equipment according to the type of equipment you are using and the manufacturer's instructions.

WARNING!
Point the torch away from your body when lighting it. Use only a spark lighter to light the torch. Do not use a match, cigarette lighter, or cigarette to light the torch.

Step 4 Heat the tubing first, and then move the flame onto the fitting. *Figure 8* shows heating the tubing and fitting.

CAUTION
The inner cone of the flame should barely touch the metal being heated. Do not direct the flame into the socket because this will burn the flux. Be sure to keep the flame moving on the metals instead of holding the flame in one place.

Step 5 Move the flame away from the joint.

Step 6 Touch the end of the solder to the area between the fitting and the tube. The solder will be drawn into the joint by the capillary action. The solder can be fed upward or downward into the joint. If the solder does not melt on contact with the joint, remove the solder and heat the joint again. Do not melt the solder with the flame.

Step 7 Continue to feed the solder into the joint until a ring of solder appears around the joint, indicating that the joint is filled. On ¾-inch-diameter tubing and smaller, the solder can be fed into the joint from one point. On larger tubing, the solder should be applied from the 6-o'clock position to the 12-o'clock position on the tubing. Generally, the amount of solder used is equal to the diameter of the tubing. For example, with ¾-inch tubing, ¾ inch of solder will fill the joint.

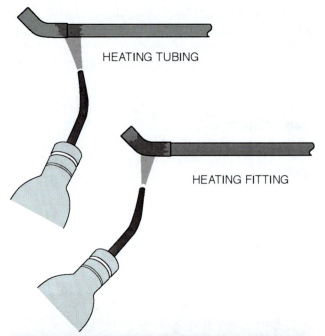

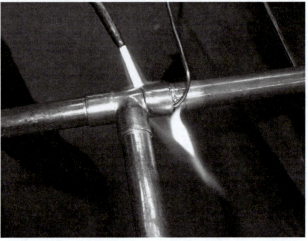

Figure 8 ◆ Heating tubing and fitting.

Step 8 Allow the joint to cool. Water can be applied to the joint to speed the cooling process.

Step 9 Wipe the joint clean with a cloth after the joint has cooled.

3.2.0 Brazing Copper Tubing and Fittings

Brazing, like soldering, uses nonferrous filler metals to join base metals that have a melting point above that of the filler metals. Brazing is performed above 800°F. Pipefitters braze tubing and fittings used in saltwater pipelines, low-pressure steam lines, refrigeration lines, medical gas lines, compressed air lines, vacuum lines, fuel lines, or other chemical lines that need extra corrosion resistance in the piping joints. Cast-bronze fittings are commonly used to connect tubings of copper, brass, copper-nickel, and steel.

Brazing produces mechanically strong, pressure-resistant joints. The strength of a brazing joint results from the ability of the filler metal to penetrate the base metal. However, penetration can only occur if the base metals are properly cleaned, the proper flux and filler metal are selected, and the clearance gap between the outside of the tubing and the inside of the fitting is 0.003 to 0.004 inch.

To properly braze copper tubing and fittings, you must understand the following materials and procedures:

- Filler metals and fluxes
- Preparing tubing and fittings for brazing
- Brazing joints

3.2.1 Filler Metals and Fluxes

Filler metals used to join copper tubing are of two groups: alloys that contain 30 percent to 60 percent silver (the BAg series) and copper alloys that contain phosphorous (the BCuP series). *Table 3* lists brazing filler materials.

WARNING!
BAg-1 and BAg-2 contain cadmium. Heating when brazing can produce highly toxic fumes. Use adequate ventilation, and avoid breathing the fumes.

The two groups of filler metal differ in their melting, fluxing, and flowing characteristics. These characteristics should be considered when selecting a filler metal. When joining copper tubing, any of these filler metals can be used; however, the most often used filler metals for close tolerances are BCuP-3 and BCuP-4. BCuP-5 is used where close tolerances cannot be held, and BAg-1 is used as a general purpose filler metal.

Brazing fluxes are applied using the same methods and rules as soldering fluxes. Brazing fluxes are more corrosive than soldering fluxes, so care must be taken never to mix a soldering flux with a brazing flux. For best results, use the flux recommended by the manufacturer of the brazing filler metals. Fluxes best suited for copper tube joining should meet *American Welding Society (AWS) Standard A5.31, Type FB3-A* or *FB3-C*.

When copper tubing is joined to wrought copper fittings with copper-phosphorus alloys (BCuP series), flux can be omitted because the copper-phosphorus alloys are self-fluxing on copper. However, fluxes are required for joining all cast fittings.

3.2.2 Preparing Tubing and Fittings for Brazing

To prepare tubing and fittings for brazing, you must follow the same procedures as you do to prepare tubing and fittings for soldering. It is critical that proper cleaning techniques be used in order to produce a solid, leakproof joint. Follow these steps to prepare the tubing and fittings for brazing.

Step 1 Measure the distance between the faces of the two fittings.

Table 3 Brazing Filler Materials

AWS Classification	Percent of Principal Element					
	Silver	Phosphorus	Zinc	Cadmium	Tin	Copper
BCuP-2	–	7–7.50	–	–	–	Balance
BCuP-3	4.75–5.25	5.75–6.25	–	–	–	Balance
BCuP-4	5.75–6.25	7–7.50	–	–	–	Balance
BCuP-5	14.5–15.50	4.75–5.25	–	–	–	Balance
BAg-1	44–46	–	14–18	23–25	–	14–16
BAg-2	34–36	–	19–23	17–19	–	25–27
BAg-5	44–46	–	23–27	–	–	29–31
BAg-7	55–57	–	15–19	–	4.5–5.5	21–23

Step 2 Determine the cup depth engagement of each of the fittings. Cup depths for fittings used with brazing are shorter than the cup depths of fittings used with soldering. The reason for the shorter cup depths is that less penetration is needed for brazing than with soldering. This measurement can be found by measuring the fitting or by using a manufacturer's makeup chart. *Figure 9* shows a manufacturer's brazing fitting makeup chart.

Step 3 Add the cup depth engagement of both fittings to the measurement found in Step 1 to find the length of tubing needed.

Step 4 Cut the copper tubing to the correct length using a tubing cutter.

Step 5 Ream the inside and outside of both ends of the copper tubing using a reamer.

Step 6 Clean the tubing and the fitting using No. 00 steel wool, emery cloth, or a special copper-cleaning tool.

CAUTION

Care must be taken when cleaning the tubing and fittings to remove all the abrasions on the copper without removing a large amount of metal. The abrasions can weaken or ruin a copper joint. Do not touch or brush away filings from the tube or fitting with your fingers because your fingers will also contaminate the freshly cleaned metal.

Step 7 Apply flux to the copper tubing and to the inside of the copper fitting socket immediately after cleaning them.

Step 8 Insert the tube into the fitting socket, and push and turn the tube into the socket until the tube touches the inside shoulder of the fitting.

Step 9 Wipe away any excess flux from the joint.

Step 10 Check the tube and fitting for proper alignment before brazing.

3.2.3 Heating Equipment

The brazing heating procedure differs from soldering in that different equipment is required to raise the temperature of the metals to be joined above 800°F. Oxyacetylene brazing equipment is used for heating brazed joints. The flame is produced by burning acetylene, which is a fuel gas, mixed with pure oxygen. *Figure 10* shows oxyacetylene brazing equipment.

MAPP gas is a mixture of liquefied petroleum and methylacetylene-propadiene, and a trademark name of Dow Chemical. Due to its combustion temperature of 5,301°F in oxygen, it is considered a much better choice for brazing than propane or other propane-based bottled gases. However, it has a lower heat value than oxyacetylene, and offers less flexibility in flame control. Generally used for smaller tubing only, it is readily available in small containers (1#) for hobbyists and other small projects. It is safer to store and transport than acetylene, and significantly safer at

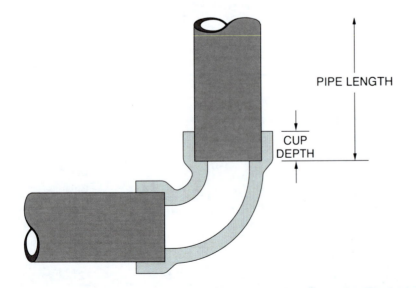

Figure 9 ◆ Manufacturer's brazing fitting makeup chart.

PIPE SIZE (in)	CUP DEPTH (in)
1/4	17/64
3/8	5/16
1/2	3/8
3/4	13/32
1	7/16
1 1/4	1/2
1 1/2	5/8
2	21/32
2 1/2	25/32
3	53/64
3 1/2	7/8
4	29/32
5	1
6	1 7/64
7	1 7/32
8	1 5/16

higher operating pressures such as those required for underwater welding applications.

Brazing is done at temperatures above 800°F. When setting the torch up, set the oxygen at about 40 psi and do not use the trigger on the torch, if it has one.

> **CAUTION**
>
> Do not run the acetylene regulator above 15 psi. At this pressure, acetylene can become unstable when mixed with air.

3.2.4 Brazing Joints

Follow these steps to braze a joint.

Step 1 Set up the oxyacetylene brazing equipment.

Step 2 Ensure that the tubing to be brazed is properly supported.

Step 3 Ensure that the tubing and fitting are properly aligned.

Step 4 Apply heat to the tubing about 1 inch from the joint, and observe the flux. The flux will bubble and turn white and then melt into a clear liquid.

Step 5 Move the heat to the fitting and hold it there until the flux on the fitting turns clear.

Step 6 Continue to move the heat back and forth over the tubing and the fitting. Allow the fitting to receive more heat than the tubing by pausing at the fitting as you continue to move the flame back and forth. When heating tubing that is 1½ inches in diameter or larger, move the heat around the entire circumference of the joint so that the entire joint is heated to brazing temperature.

Step 7 Touch the filler metal rod to the joint when a puddle of molten copper forms at the joint.

> **NOTE**
>
> If the filler metal melts upon contact with the joint, the brazing temperature has been met, and you should proceed to Step 8. If the filler metal does not melt upon contact, continue to heat and test the joint until the filler metal melts.

Step 8 Hold the filler metal rod to the joint, and allow the filler metal to enter into the joint while holding the torch slightly ahead of the filler metal and directing most of the heat to the shoulder of the fitting.

Step 9 Continue to fill the joint with the filler metal until the filler metal inside the joint fills to a depth of at least three times the thickness of the tubing. This will not fill the joint, but a joint with this amount of penetration will be stronger than the tubing. If the tubing is 2 inches or more in diameter, 2 torches can be used to evenly distribute the heat. For larger joints, small sections of the joint can be heated and brazed. Be sure to overlap the previously brazed section as you continue around the fitting. *Figure 11* shows working in overlapping sectors.

Step 10 Wash the joint with warm water to clean excess, dried, or hardened brazing flux from the joint. If the flux is too hard to be removed with water, chip the excess flux off using a small chisel and light ball-peen hammer, and then wash the joint with warm water. Joints clean best when they are still warm.

Step 11 Allow the joints to cool naturally.

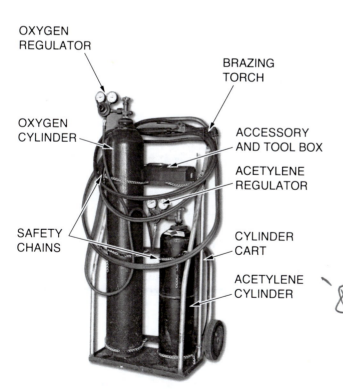

NOTE: Tanks may not be stored on this carrier except when in use.

Figure 10 ♦ Oxyacetylene brazing equipment.

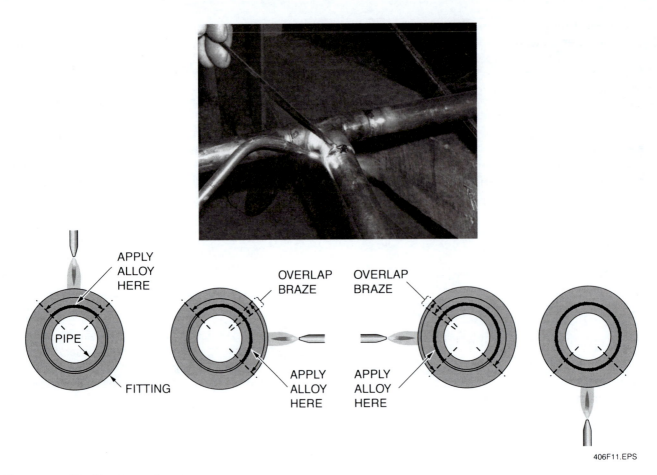

Figure 11 ♦ Working in overlapping sectors.

4.0.0 ♦ BENDING PIPE

When designing a piping system, an engineer has two choices to make when determining how to change directions of the piping system. One is to buy fittings (butt weld, socket weld, or threaded) to change the directions of the piping system, and the other is to use pipe benders to bend the pipe. Bending the pipe allows the process to flow more freely through the system without restrictions caused by fittings. *Figure 12* shows a 90-degree screw fitting and a 90-degree bend.

Pipe bending can be accomplished by either cold or hot bending methods. Some of the factors that influence the pipe bending process are the size of the pipe, the material composition of the pipe, and the number of identical bends required for the piping system.

Both the cold and hot bending methods can cause distortion to occur. This happens because the outside of the bend will stretch until the pipe flattens or collapses, causing the inside of the pipe to buckle or wrinkle. To prevent this distortion, dry sand or a mandrel is often placed inside the pipe during the bending process. *Figure 13* shows different types of bends.

4.1.0 Calculating Pipe Bends

There are tables that give the dimensions for commonly used bends, but these tables do not provide information for all possible bends. If the information for the fabrication or replacement of bends is not available in tables or on the fabrication drawings, the pipefitter must know how to calculate the pipe bend. Most formulas give the angle of the bend, which is normally referred to as the degree of the bend (D). To calculate pipe bends, you must recognize a few pipe bending terms (see *Figure 14*).

Pipefitters must know how to calculate the following:

- Bend allowance
- Angle of bend
- Radius of bend

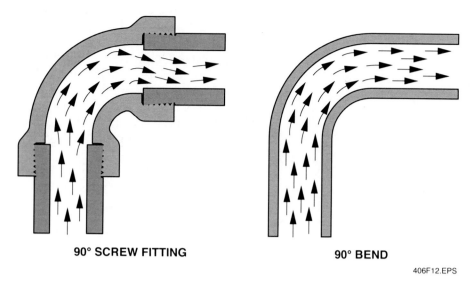

Figure 12 ◆ 90-degree screw fitting and 90-degree bend.

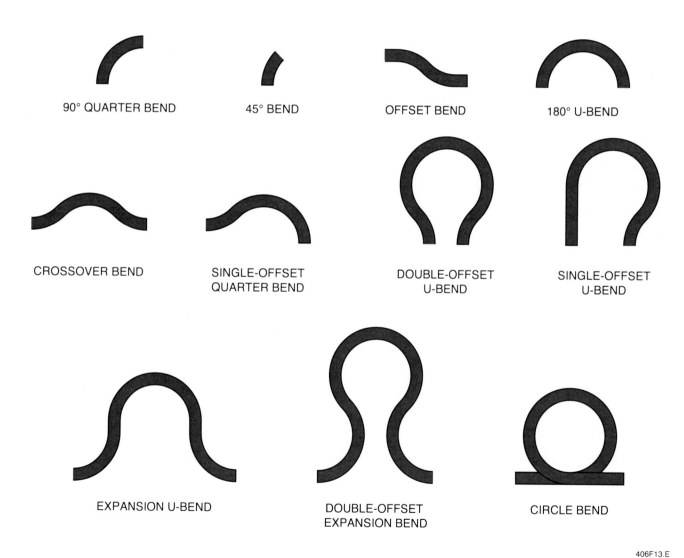

Figure 13 ◆ Types of bends.

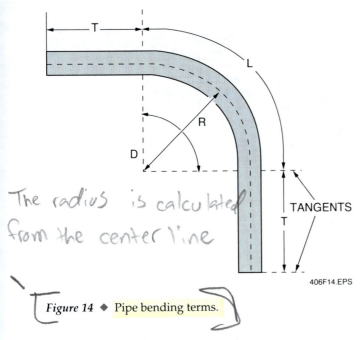

Figure 14 ♦ Pipe bending terms.

4.1.1 Calculating Bend Allowance

The length of straight pipe needed to make a curved bend is called bend allowance. The pipefitter can determine how much straight pipe is needed by using the following formula:

L = R × D × 0.01745

L = Length of pipe needed for bend

R = Radius of bend

D = Number of degrees in bend

For example, if you are making a 90-degree bend, D is 90, so L = R × 90 × 0.01745.

Remember that large pipe must be bent with a long radius. To prevent the pipe from overstretching or collapsing, the radius of the bend should be five times the diameter of the pipe. The radius may be larger but should not be smaller. According to this rule, if you are bending 2-inch pipe, the bend radius should be 5 × 2 inches, or 10 inches. If the bend radius is 10 inches, R is 10, and L = 10 × 90 × 0.01745.

Once R and D are known, L can be found by multiplying as follows:

L = 10 × 90 × 0.01745

10 × 90 = 900

L = 900 × 0.01745

900 × 0.01745 = 15.705

L = 15.705

According to the formula, the length of pipe needed for the bend, or the bend allowance, is 15.705 inches.

To find the total length of a pipeline containing a bend, add the bend allowance to the length of straight pipe on each side of the bend. These straight lengths are called tangents. In *Figure 15*, the total length of the line is 27.705 inches since 7" + 5" + 15.705" = 27.705".

Follow these steps to calculate the bend allowance and total length of line needed for a 70-degree bend in 1-inch pipe (see *Figure 16*).

Step 1   Write down the bend allowance formula.

L = R × D × 0.01745

Step 2   Find D.

L = R × 70 × 0.01745

Step 3   Calculate R.

1" pipe × 5 = 5"

L = 5 × 70 × 0.01745

Step 4   Multiply R × D.

5 × 70 = 350

Step 5   Multiply the total by 0.01745.

350 × 0.01745 = 6.1075

The bend allowance (L) = 6.1075

Step 6   Add both tangents.

15" + 10" = 25"

Step 7   Add the length of both tangents to the bend allowance.

6.1075" + 25" = 31.1075"

The total length of line needed for the run is 31.1075 inches.

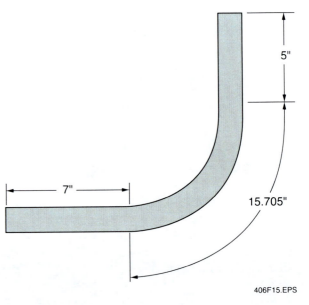

Figure 15 ♦ Determining line length.

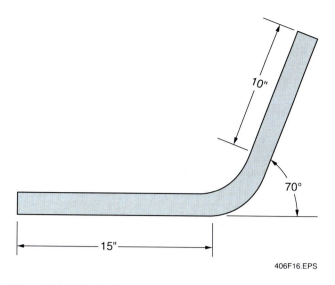

Figure 16 ◆ Bent pipe run.

4.1.2 Calculating Angle of Bend

A pipefitter must sometimes replace a bend in a pipe run. The new bend must often be the same size and angle as the old bend. If drawings showing the angle and radius of the bend are not available, the pipefitter can find the angle of the bend, using two straightedges and a protractor. *Figure 17* shows how to set up the straightedges and protractor to find the bend angle.

The two straightedges are placed on the tangents of the bend, which are the straight pieces of the pipe run. The center point of the protractor is then placed at the point where the two straightedges cross. This point, called the vertex, is the point where two straight lines meet to form an angle. *Figure 18* shows two examples of a vertex.

When the protractor is lined up on one straightedge and the center point is placed at the vertex, the angle is easy to read. Mark the point where the second straightedge crosses the dial of the protractor. This mark is the angle of the bend. *Figure 19* shows a protractor.

4.1.3 Calculating Radius of Bend

The easiest way to find the radius of a pipe bend is to use a square and rule. Place the end of the square on one tangent at the point where the bend begins. Draw a line from point A to point B as shown in *Figure 20*. Repeat these steps for the other tangent, and draw a line from point C to point D. Measure the distance from point E to where the two lines cross to the center of the pipe bend. This distance is the radius of the bend. *Figure 20* shows finding the radius.

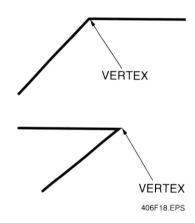

Figure 18 ◆ Vertex.

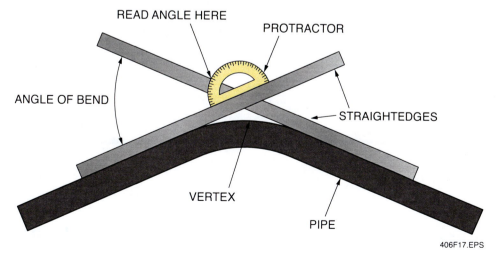

Figure 17 ◆ Finding bend angle.

4.2.0 Laying Out Bends

Dimensions of the pipeline are usually included on the pipe drawings. The dimensions are given from the center lines of the pipe. The radius of a bend is given from the center line of the pipe. Once the radius dimension and the degree of the angle of the bend are known, you can figure the layout and the length of pipe that you must cut in order to bend the pipe correctly. To lay out a pipe bend, you need a protractor and the blade from a combination set. Remember that you have to measure the inside radius of the bend and that the calculations need to account for the radius. Follow these steps to lay out a 45-degree bend on a 4-inch pipe with a 36-inch radius.

Step 1 Subtract one-half of the OD of the 4-inch pipe from the 36-inch radius center line to obtain the inside radius of the bend.

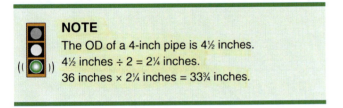

NOTE
The OD of a 4-inch pipe is 4½ inches.
4½ inches ÷ 2 = 2¼ inches.
36 inches × 2¼ inches = 33¾ inches.

Step 2 Draw an arc with a radius of 33¾ inches (*Figure 21*).

Step 3 Draw a straight line to meet the arc, and label the line AB (*Figure 22*). Line AB is the tangent of the arc, which means it touches the arc in one place.

Step 4 Lay out a 45-degree angle from line AB, using the combination set protractor and blade.

Step 5 Set the protractor at 45 degrees, and slide the protractor along line AB until the blade meets the arc.

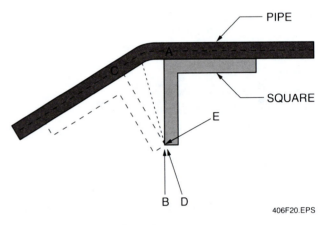

Figure 20 ◆ Finding the radius.

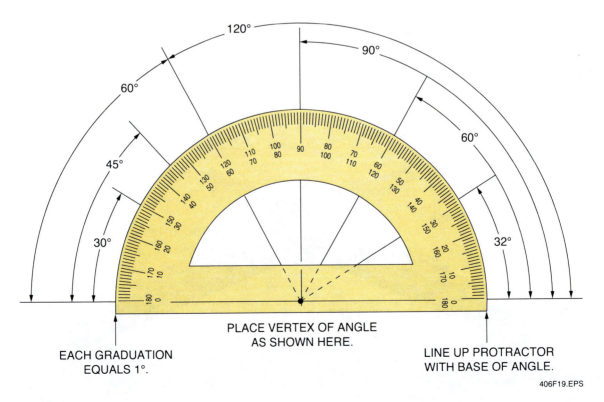

Figure 19 ◆ Protractor.

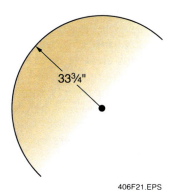

Figure 21 ◆ Arc with radius of 33¾ inches.

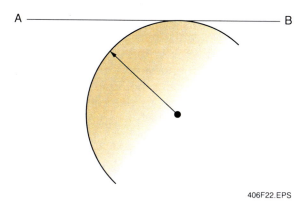

Figure 22 ◆ Line AB.

does soften the metal. When hot-bending, extra steps must be taken to prevent the pipe walls from collapsing.

Pipefitters use the following methods to bend pipe:

- Hydraulic bending
- Manual bending

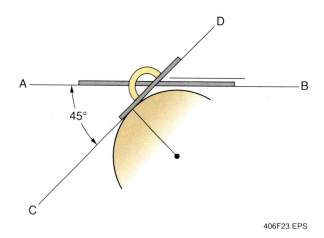

Figure 23 ◆ Line CD.

Step 6 Draw in the layout of the bend along the straight line of the protractor, and label this line CD as shown in *Figure 23*.

Step 7 Measure 2¼ inches (½ OD) at right angles to lines AB and CD to draw the center lines (*Figure 24*).

Step 8 Draw extension lines at right angles from the center of the bend on each center line and another extension line at each end of the bend. *Figure 25* shows the length of the bend (LB).

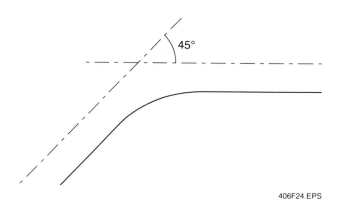

Figure 24 ◆ Center lines.

4.3.0 Methods of Pipe Bending

There are different kinds of benders used to either cold-bend or hot-bend pipe. Bending cold pipe has several advantages. It requires less time because no preheating or special handling is needed. Also, cold-bending does not anneal the pipe and soften it. Cold-bending produces a stiffer, more rigid pipe. Pipes can be cold-bent using either the hand-bending or machine-bending methods.

It is sometimes necessary to heat thick wall piping and tubing before bending. Hot-bending requires less force than cold-bending, but heating

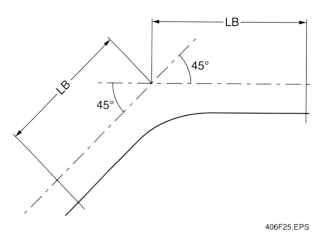

Figure 25 ◆ Layout of LB.

4.3.1 Hydraulic Bending

The hydraulic bender is a power-driven tool that uses a ram that fits into the bending frame. The hydraulic bender is one of the easiest benders to use. Most hydraulic benders have a bending shoe that fits over the head of the ram. Different size pipe requires a different size bending shoe and swivel shoes that fit into the holes for the nominal pipe size to be bent. An angle gauge is connected to the swivel shoe to indicate the angle of the bend.

To use a hydraulic pipe bender, the pipe must be inserted at the proper position where the bend is needed. Make sure your hands are away from the hydraulic ram, and then activate the ram until the stroke completes the bend. Release the hydraulic ram and remove the pipe. *Figure 26* shows a typical hydraulic bender. Follow manufacturer's instructions on these and any other machines.

Pipe bending may be a task required only a few times on a given project, depending upon the craft. In those cases, manual or light-duty hydraulic benders would be the first choice. Most light-duty, highly portable hydraulic benders utilize a compact hand pump to provide the hydraulic pressure, although electric hydraulic pumps are available that can increase speed and productivity. However, some projects may require hundreds, or even thousands, of bends. For projects requiring this level of production, heavy-duty hydraulic or electric benders become the tool of choice. They offer much higher levels of production, improved consistency for repetitive bends, and reduced operator fatigue.

4.3.2 Manual Bending

Manual benders are normally used on EMT (electro-metallic tubing) electrical conduit and on 1" or smaller rigid pipe, due to the obvious manual effort required to work with larger sizes. Typical benders feature a foot pedal to provide the necessary leverage for bending; no other equipment is required. Hickey benders are designed exclusively for bending rigid pipe, with the pipe placed in a sturdy vise. The head of each type has an internal radius that is shaped to fit the contour of the pipe, with an arm or hook to pull the pipe around the radius during bending. Manual benders are generally constructed of aluminum or steel, while Hickey benders are usually constructed of steel or iron. *Figure 27* illustrates both types of benders.

>
> **NOTE**
> The following procedure explains how to use a Hickey bender to bend a 1-inch pipe that is 12 inches long to a 90-degree bend, starting 8 inches from the end-to-center after the bend is made.

Hickey benders are made in different sizes to fit the standard pipe diameters. Follow these steps to bend pipe manually using a Hickey bender.

Step 1 Measure from the end of the pipe 8 inches, and mark the pipe at this point. This is measurement A (see *Figure 28*).

Step 2 Calculate the radius of the pipe.
R = 1 inch × 5 = 5 inches

Step 3 Measure back 5¾ inches along the pipe from measurement A, and label the radius on the pipe as B (*Figure 29*). This is the point where the bend or arc begins.

Step 4 Calculate the length of the bend (LB). LB = 9 inches. This is the distance from B past A to C. *Figure 30* shows measurement BC.

MODEL 3 TUBING BENDER

MODEL 3 TUBING BENDER (SIDE VIEW)

Figure 26 ◆ Hydraulic and electric benders.

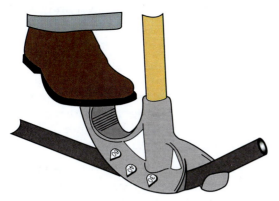

THIN-WALL CONDUIT BENDER

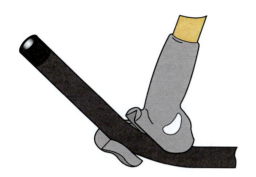

PIPE AND HEAVY-WALL CONDUIT BENDER

Figure 27 ◆ Hand benders.

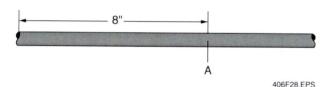

Figure 28 ◆ Measurement A.

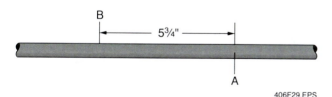

Figure 29 ◆ Marking radius on pipe.

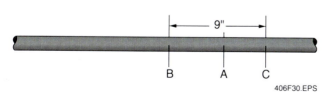

Figure 30 ◆ Measurement BC.

Step 5 Place the pipe in a vise at point B.
Step 6 Place the Hickey bender at point C.
Step 7 Bend the pipe 90 degrees.

5.0.0 ◆ GLASS-LINED PIPING

Glass-lined pipe refers to cast iron, carbon steel, stainless steel, and alloy pipe that is lined with a specified grade and thickness of glass. Carbon steel is the most widely used for glass lining because of its relatively low cost, its availability, and its ease of fabricating and glassing. Glass can only be applied to smooth, accessible surfaces. Machined surfaces cannot be glass-lined. Glass-lined piping is widely used in many chemical and food-processing industries because of the following advantages:

- *Cleanliness* – Glass does not contaminate products, and it protects the color and purity of products.
- *Smoothness* – The glass lining minimizes friction, thus reducing wear on agitating equipment. The smoothness of the glass also makes it resistant to viscous or sticky product buildup.
- *Superior vacuum service* – Although glass linings are comparable in pressure and temperature ratings to other linings, the tight bond between the glass and steel provides superior vacuum service.
- *Corrosion resistance* – Glass resists corrosion in most acid solutions, with the exception of hydrofluoric acid, and resists corrosion in alkali solutions at moderate temperatures.

The disadvantages of glass piping are that the cost is very high, it is very fragile, and it can be very troublesome to work with.

Three of the major suppliers of glass-lined pipe in North America include Pfaudler, De Dietrich, and The Ceramic Coating Company. Engineers at these companies can recommend and specify glass formulations and thicknesses for particular services. These companies also offer specific training on the installation and repair of glass-lined piping and equipment.

This module gives a general overview of removing and installing glass-lined piping, performing preventive maintenance on glass-lined piping, and making shims for large-diameter gaskets used with flanged glass piping systems.

5.1.0 Removing and Installing Glass-Lined Piping

Follow these steps to remove and install a section of glass-lined piping.

Step 1   Drain the pipe section to be removed.

Step 2   Valve out, tag, and lock the pipe section to be removed.

Step 3   Support the pipe section.

Step 4   Loosen and remove the nuts and bolts from the first flange joint.

Step 5   Remove the old gasket from the first flange joint. Record the size and type of the gasket before discarding it.

Step 6   Loosen and remove the nuts and bolts from the second flange joint.

Step 7   Remove the old gasket from the second flange joint. Record the size and type of the gasket before discarding it.

Step 8   Disconnect the pipe section from all pipe supports present.

Step 9   Remove the old pipe section from the pipeline.

CAUTION

Never strike glass-lined pipe and fittings with tools or other hard objects because damage could result. Mechanical stresses caused by unequal or excessive bolt torque, springing or pulling the pipe into position, or clamping the pipe too tightly (which prevents uniform thermal expansion) can result in damage to the glass.

Step 10   Remove the old pipe section from the restraining device.

Step 11   Obtain the new pipe section.

Step 12   Install the new pipe section into the restraining device.

Step 13   Obtain the new gaskets.

NOTE

Self-centering Teflon® gaskets are often recommended. However, check specifications for the product in use to ensure the appropriate gasket is used.

Step 14   Depending on application and/or specification, apply anti-seize compound to the hardware.

Step 15   Ensure that the gaskets match the size and type of the original gasket exactly, as recorded in Steps 5 and 7. Do not use any gaskets that are not of the correct size and type. Leakage of the product could result from using the wrong size gaskets.

Step 16   Ensure that any foreign material on the gasket, such as grease, product, or adhesive tape, is removed before installing the gasket.

Step 17   Install the pipe into position, but do not make the flange joint connections.

CAUTION

Never pull or spring a pipe section into place. This could cause the pipe to warp, causing an improper fit. Never weld to the exterior of a glass-lined pipe because the heat from welding may separate the glass lining from the pipe.

Step 18   Ensure that all joints are on the same center line, with the flanges nearly touching (normally within ¼ inch and never more than ½ inch apart).

Step 19   Install the pipe section into the pipe hangers.

Step 20   Ensure that the clamping ring used in the pipe hanger is loose enough around the outside surface of the pipe to allow for the thermal stress.

NOTE

On vertical runs, use riser clamps above or below floors. Where possible, install the clamp below a joint. Do not support the piping too rigidly. Allow for some freedom of movement to relieve thermal stress.

Step 21   Ensure that the pipe supports are positioned at 10-foot intervals along the length of the pipe to keep the joints from sagging.

Step 22   Measure the distance between the flanges at several points around the circumference, and adjust the piping or shim the gaskets to correct misalignment between the two flanges.

Step 23 Check the alignment of the flanges.

Step 24 Install the gaskets into position in the flange joints. Be sure to keep the gaskets dry during installation. Depending on the material, wet gaskets may slip under pressure.

Step 25 Ensure that the position of the gasket allows the entire flange seating area to be covered by the gasket.

Step 26 Ensure that all joints, including the mating flanges, are parallel and have enough freedom of movement to permit uniform tightening and to avoid stress when the flange joints are connected.

Step 27 Install the nuts and bolts into the first flange joint, tightening the bolts by hand at first.

Step 28 Tighten two bolts that are 180 degrees apart, and then advance in the bolt-tightening sequence shown in *Figure 31*. Ensure that the bolt is held in place when the nut is tightened. To prevent twisting the gasket out of alignment, be sure not to rotate the bolts. Always use a torque wrench to tighten the bolts.

Step 29 Tighten the bolts to 25 percent of the recommended torque value, using a torque wrench. *Table 4* lists some recommended bolt torques.

> **NOTE**
> Ensure that nuts, bolts, and lubrication comply with the job specifications.

Step 30 Continue to tighten the bolts in increments of 25 percent of the recommended torque value, using the crossover tightening method, until the recommended torque value is reached.

Step 31 Install the nuts and bolts onto the second flange joint, tightening the bolts by hand at first.

Step 32 Repeat Steps 28 through 30 to tighten the bolts to the correct torque.

Step 33 Retighten all bolts and clamps after one operating cycle or after at least 24 hours at rest.

Step 34 Remove all locks and tags.

Step 35 Check the flanges for leaks once the pipeline is back on line and full of product.

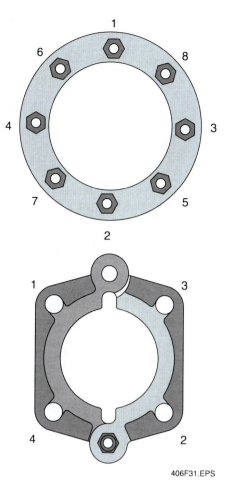

Figure 31 ♦ Bolt-tightening sequence.

5.2.0 Performing Preventive Maintenance on Glass-Lined Piping

Perform the following preventive maintenance procedures during routine inspections of the pipeline:

- Check bolt torques at all joints regularly.
- Check all joints for leaks. If a joint leaks, do the following:
  - Make sure the bolts are tightened evenly.
  - Replace the gasket if necessary.
  - Correct any possible cause of strain at the joint.
- Watch for acid spills on the outside of the pipe or fitting. Hose off any spills immediately. Acid spills may react with the steel, causing chipping or fish-scaling of the glass on the inside.
- Check periodically for any external damage that may have caused breakage of the glass on the inside. Be sure to replace any damaged part.

MODULE 08406-07 ♦ SPECIAL PIPING 6.23

Table 4 Torque Specifications

TORQUE IN FOOT POUNDS

Fastener Diameter	Threads Per Inch	Mild Steel	Stainless Steel 18-8	Alloy Steel
1/4	20	4	6	8
5/16	18	8	11	16
3/8	16	12	18	24
7/16	14	20	32	40
1/2	13	30	43	60
5/8	11	60	92	120
3/4	10	100	128	200
7/8	9	160	180	320
1	8	245	285	490

SUGGESTED TORQUE VALUES FOR GRADED STEEL BOLTS

Grade	SAE 1 or 2	SAE 5	SAE 6	SAE 8
Tensile Strength	64,000 PSI	105,000 PSI	130,000 PSI	150,000 PSI

Bolt Diameter	Threads Per Inch	Foot Pounds Torque			
		SAE 1 or 2	SAE 5	SAE 6	SAE 8
1/4	20	5	7	10	10
5/16	18	9	14	19	22
3/8	16	15	25	34	37
7/16	14	24	40	55	60
1/2	13	37	60	85	92
9/16	12	53	88	120	132
5/8	11	74	120	169	180
3/4	10	120	200	280	296
7/8	9	190	302	440	473
1	8	282	466	660	714

- Inspect all gasketed joints at regular intervals for weeping, leaks, or deterioration. Because a gasket is subjected to temperature and pressure cycling, material relaxation, and corrosive environments, an initial 3-month inspection is recommended. The bolt torque should also be checked at this time.

6.0.0 ◆ HYDRAULIC FITTED COMPRESSION JOINTS

Hydraulic compression joints offer significant advantages over many other types of pipe joining techniques in many applications. They can be used on small or large projects, where flammable or explosive conditions exist which prohibit an open flame, and even under water. A number of companies manufacture a complete line of hydraulic compression joints. Two popular product lines are those manufactured by Lokring® and Ridgid's ProPress® System. Installation of both of these products is covered in this module.

6.1.0 Lokring® Hydraulic Compression System

The Lokring® hydraulic compression joint is designed with a patented elastic strain preload (ESP®). The Lokring® fitting has a precision-machined metallic fitting body and two swaging rings that are factory-assembled onto either end of the fitting body. The patented design has no O-ring seals, threads, or ferrules and is used on pipe or tubing without heat or threading. *Figure 32* shows the Lokring® fitting components.

Pipe is inserted into the ends of the fitting body, and the hydraulic installation tool is engaged onto the fitting. As the hydraulic pressure is actuated on the tool, the tool advances each swage ring axially over the fitting body until the leading edge of the swage ring contacts the fitting body tool flange. *Figure 33* shows the movement of the Lokring® fitting.

As the swaging rings pass over the tapered section of the fitting body they move over a slightly increased diameter, compressing the fitting body down onto the pipe surface. The fitting body compressed onto the pipe causes the sealing lands of the fitting body to grip and seal on the pipe surface, resulting in a permanent, gas-tight, metal-to-metal seal between the pipe surface and the fitting body.

The Lokring® fittings require a Loktool® installation tool, which is designed to work with a single-acting hydraulic pump rated at 10,000 psi operating pressure. The following general safety instructions apply to the use of high-pressure hydraulic systems, and should be adhered to at all times:

- Verify that all hydraulic connection hoses and fittings are rated for a minimum of 10,000 psi.
- Do not overtighten the hydraulic system threaded connections.
- Purge the air from the hydraulic system.
- Use only hydraulic pumps, hoses, and fittings supplied or recommended by Lokring®.
- Never disassemble the Loktool® installation tool.
- Ensure that the hydraulic hose is fully engaged on the fitting before actuating the hydraulic pressure.

1. GAUGE AND MARK PIPE ENDS.
2. POSITION FITTINGS OVER MARKS.

3. ENGAGE LOKTOOL® HEAD.
4. MAKE UP SWAGE RING.

5. REMOVE LOKTOOL® HEAD. INSPECT COMPLETED JOB.

Figure 33 ♦ Movement of Lokring® fitting.

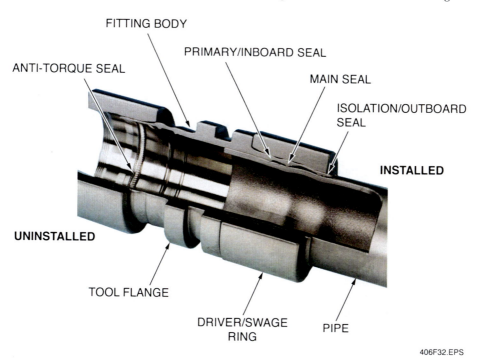

Figure 32 ♦ Lokring® fitting components.

- Release the hydraulic power when the fitting swaging rings seat against the fitting center stop.
- Do not drop heavy objects on the hydraulic hoses.
- Do not use hydraulic hoses that are kinked or sharply bent.
- Keep hydraulic equipment and hoses away from flames and heat.
- Always provide adequate clearance between hoses and couplers to avoid moving objects that may cause abrasion or cuts.

6.2.0 Installing Lokring® Fittings

Refer to the Lokring Technology's document *LP-101* or *LP-105 Installation Procedure for Lokring® Fittings* when installing Lokring® fittings. Follow these steps to install Lokring® fittings dry:

Step 1 Obtain 120 grit aluminum oxide cloth (required). Also obtain 60 grit cloth to remove deep pits or scratches.

Step 2 Obtain Loctite Compound PST 567 (pipe sealant with Teflon®). This compound may be applied as an aid to seal where the pipe surface is somewhat poor, and is suggested for all applications using Schedule 10 pipe or thinner.

Step 3 Obtain a Tri-Tool pipe facing tool or equivalent (optional).

Step 4 Cut the pipe to the desired length, using a pipe cutter that will provide a square cut and will not flatten or deform the pipe end. A square cut with a maximum of 5 degrees off square is required.

Step 5 Remove the inside and outside burrs, making sure to use good pipe handling techniques. For fit-ups with no gaps between the pipe ends, a Tri-Tool pipe facing tool should be used.

Step 6 Sand the pipe ends manually using fine grit (120#) aluminum oxide cloth around the pipe end for a minimum of 1½ pipe diameters from the end of the pipe.

CAUTION

Always sand around the circumference of the pipe. Sanding along the axis of the pipe can result in undesirable flat spots on the pipe end surface. Extra care must be taken to avoid gouging or flattening the pipe surface during sanding.

NOTE

If deep pits or scratches persist, use 60 grit aluminum oxide cloth to remove, followed by the 120 grit cloth for final finish.

Step 7 Inspect the cleaned pipe end in a well-lighted area, and wipe the pipe end clean to remove abrasive residue.

Step 8 Obtain a Lokring® NO GO gauge.

Step 9 Attempt to pass the sanded ends of the pipe through the end of the NO GO gauge at two points 90 degrees apart. If the pipe passes through at either point, it is too small and should not be used. *Figure 34* shows using a NO GO gauge.

Step 10 Slide the larger (hex) end of the NO GO gauge over the pipe end until the gauge bottoms out on the pipe end. If the pipe does not fit into the NO GO gauge, the pipe is oversized and should not be used.

Step 11 Mark the pipe through the slots on the gauge, using a permanent marker as shown in *Figure 35*. Each pipe end must be marked so that the quality of the pipe installation can be verified.

PIPE CORRECTLY SIZED
(PIPE DOES NOT PASS THROUGH GAUGE)

PIPE UNDERSIZED
(PIPE PASSES THROUGH GAUGE)

USED WITH PERMISSION OF LOKRING CORPORATION.

Figure 34 ◆ Testing pipe ends using NO GO gauge.

Step 12 (*Optional*) Apply Loctite PST 567 pipe sealant (*Figure 36*). Apply to the end of the pipe, just back from the end of the pipe and up to the inspection mark. Care must be taken to prevent excess sealant material from entering the pipe end and potentially contaminating the piping system.

Step 13 Place the Lokring® fitting onto one pipe end.

Step 14 Align the second pipe with the first pipe, and butt the ends together.

Step 15 Position the Lokring® fitting centrally over the two pipe ends so that the inspec-

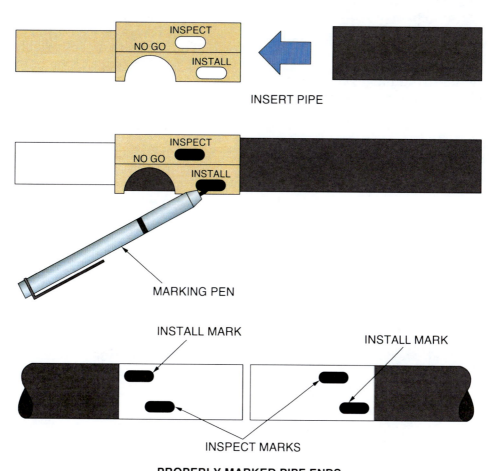

Figure 35 ◆ Marking pipe.

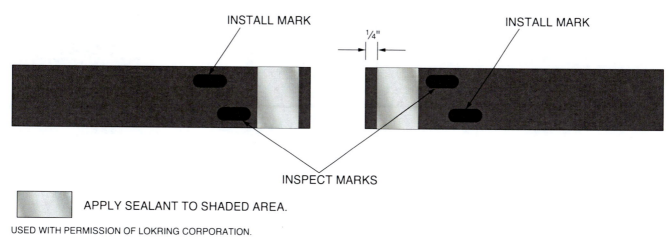

Figure 36 ◆ Application of anaerobic sealant.

tion marks on both pipes are covered and part of the installation marks are visible. *Figure 37* shows a proper alignment and fit-up of the pipe prior to assembly.

Step 16 Set up the Loktool® system components (*Figure 38*).

Step 17 Install the Loktool® head on the Lokring® fitting (*Figure 39*).

Step 18 Check the position of the fitting on the pipe end.

Step 19 Actuate the hydraulic power until the moving jaw forces the Lokring against the external center stop. *Figure 40* shows the first end being made up. The fitting makeup is complete when the drive ring contacts the tool flange.

Step 20 Remove the Loktool® from the installed fitting end.

Step 21 Inspect the Lokring® fitting. First, the outboard *Inspect* mark should be completely uncovered, but the inboard *Install* mark should be partially covered by the trailing edge of the fitting body. Second, the drive ring should contact the tool plane. Third, the end of the fitting body protrudes beyond the drive ring. *Figure 41* shows a first end installation.

Step 22 Before installing the second pipe, ensure that the pipes are properly aligned. This will avoid stressing the fitting.

Step 23 Repeat Steps 17 through 21 to prepare and install the Lokring® fitting on the joining pipe.

USED WITH PERMISSION OF LOKRING CORPORATION.

Figure 37 ◆ Proper alignment and fit-up.

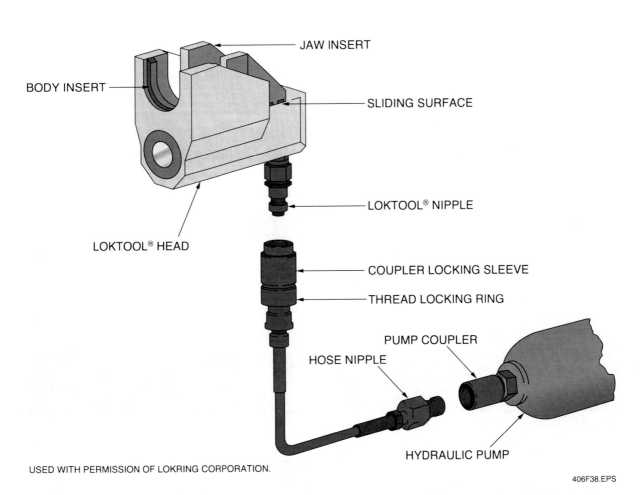

USED WITH PERMISSION OF LOKRING CORPORATION.

Figure 38 ◆ Loktool® system components.

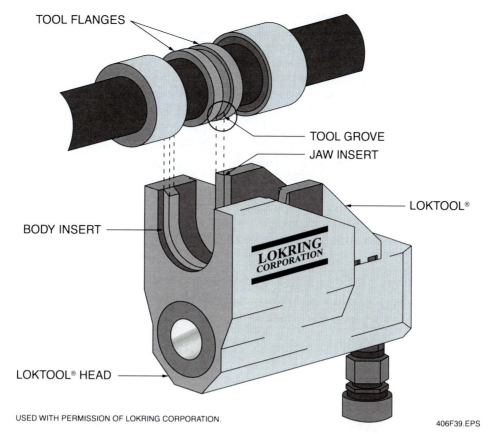

Figure 39 ♦ Loktool®.

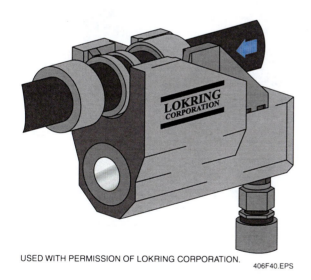

Figure 40 ♦ First end makeup.

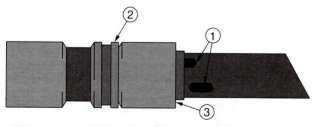

Figure 41 ♦ First end installation.

7.0.0 ♦ GROOVED PIPING SYSTEMS

Grooved piping systems provide an easy alternative to welded, flanged, or threaded joints. Each joint in a grooved piping system serves as a union, allowing easy access to any part of the piping system for cleaning or servicing. The leading manufacturer of grooved piping couplings, fittings, and fabrication tools is Victaulic Company of America. Grooved piping systems have a wide range of applications and can be used with a wide variety of piping materials. The most common uses of grooved piping systems are in oil fields and industrial facilities, and mining, municipal, and fire protection systems. The following types of piping can be joined by grooved couplings:

- Carbon steel
- Stainless steel
- Aluminum
- PVC plastic
- High-density polyethylene
- Ductile iron

The standard grooved piping coupling consists of a rubber gasket and two housing halves that are bolted together. The housing halves are tightened together until they touch, so no special torquing of the housing bolts is required.

MODULE 08406-07 ♦ SPECIAL PIPING 6.29

The grooved piping system offers varied mechanical benefits, including the option of rigid or flexible couplings. Rigid and flexible couplings can be incorporated as needed into any system to take full advantage of the characteristics of each. Rigid couplings create a rigid joint useful for risers, mechanical rooms, and other areas where positive clamping with no flexibility within the joints is desired. Flexible couplings provide allowance for controlled pipe movement that occurs with expansion, contraction, and deflection. Flexible couplings may eliminate the need for expansion joints, cold springing, or expansion loops and will provide a virtually stress-free piping system. *Figure 42* shows an example of a rigid grooved coupling and a flexible grooved coupling.

In addition the cut and rolled grooves, several companies have also developed an equivalent of the Ridgid ProPress® joining system, used with Schedule 5 and Schedule 10 pipe. The pipe is pushed together, one end inside the other, and formed into a groove with a special tool (*Figure 43*) in the field.

The ProPress® tool is a hydraulic clamp that forms the pipe to the fittings. The pipe slips over the bead, and the tool compresses the pipe over the bead to form a tight fit.

To form a ProPress® joint, perform the following steps:

Step 1 Choose the correct size jaw for the tool.

Step 2 Be sure the tool is unplugged or the battery removed. Check that the locking pin on the tool is fully open; that is, the pin is pulled out until it stops. Hold the tool over the lever end of the jaws and slide the tool into position.

Step 3 Push the locking pin into the locked position.

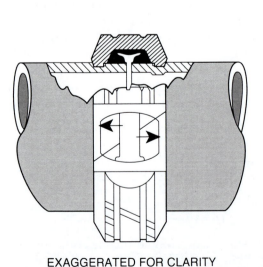

EXAGGERATED FOR CLARITY
ZERO-FLEX® RIGID COUPLING

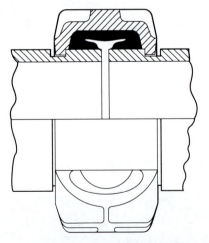

EXAGGERATED FOR CLARITY
STANDARD FLEXIBLE COUPLING

Figure 42 ◆ Rigid and flexible grooved pipe couplings.

Step 4 Grasp the tool by the handgrip and the levers of the jaws, opening the jaws. Slide the jaws over the pipe at the bead on the fitting.

Step 5 When the jaws are in place, move your hand off the levers. Press the trigger. The jaws will close fully (*Figure 44*) and retract.

CAUTION

The tool will close fully when the trigger is pulled, whether the trigger is released or not. Be very careful that no fingers are between the jaws when the trigger is pulled.

The tool jaws should be cleaned at regular intervals as instructed in the documentation supplied by the manufacturer. Read and obey all instructions in the manual.

Figure 43 ◆ ProPress® tool.

Figure 44 ◆ Holding the ProPress® tool in use.

7.1.0 Preparing Pipe Ends

Grooved pipe can be delivered to the job site precut to length and grooved, or it can be cut and grooved on the job. Use of the grooved piping method is based on the proper preparation of a groove in the pipe end to receive the coupling housing key. The groove serves as a recess in the pipe with enough depth to secure the coupling housing, yet, at the same time, leaves enough wall thickness for a full pressure rating. Groove preparation varies with different pipe materials and wall thicknesses. A variety of tools is available from Victaulic Company of America to properly groove pipe in the shop or in the field. The two methods of forming a groove in pipe include roll grooving and cut grooving. *Figure 45* shows details of pipe grooves.

The dimensions pointed out by the letters in *Figure 45* must comply with engineering specifications at your job site. The A dimension is the distance from the pipe end to the groove and provides the gasket seating area. This area must be free from indentations, projections, or roll marks to provide a leakproof sealing seat for the gasket. The B dimension is the groove width. This dimension controls expansion and angular deflection based on its distance from the pipe end and its width in relation to the width of the coupling housing key. The C dimension is the proper diameter of the base of the groove. This dimension must be within the given diameter tolerance and concentric with the outside diameter of the pipe. The D dimension is the nominal depth of the groove, and is used as a trial reference only. If necessary, the D dimension must be changed to keep the C dimension within the stated tolerances. The F dimension is used with the standard roll only, and gives the maximum allowable pipe end flare. The T dimension is the lightest grade, or minimum thickness, of pipe suitable for roll or cut grooving. The R dimension is the radius necessary at the bottom of the groove to eliminate a point of stress concentration for cast iron and PVC plastic pipe.

7.1.1 Roll Grooving

Power roll-grooving machines are used to roll grooves at the ends of pipe to prepare the piping for groove-type fittings and couplings. Power grooving machines are available to groove 2- to 16-inch lightweight steel pipe, aluminum pipe, and stainless steel pipe. *Figure 46* shows a power roll-grooving machine.

Many power roll-grooving machines are stand-alone units with their own power drives. Although some self-contained units are for field use, others are designed for shop fabrication, and work at higher productivity rates. A number of roll grooving units are manufactured without power drives, and are designed for use with the Ridgid 300 (or equal) power drive. Such drives can then be used for both pipe threading and grooving operations in the field, saving the expense of multiple drives on the same site. Even small roll grooving units, designed to be hand-cranked for small projects, offer accessories to allow the connection of a power drive.

The roll-grooving machine presses a groove into the pipe, using upper and lower die heads. The lower die head, or female die, rotates the pipe. The upper die head, or male die, is powered either by hydraulic pressure or a handwheel and pushes a groove into the pipe as the pipe is rotated.

Roll grooving removes no metal from the pipe but forms a groove by displacing the metal. Since the groove is cold-formed, it has rounded edges that reduce the pipe movement after the joint is made up. However, it is also important to note that since the grooving process literally squeezes the material toward the center, a deformity results on the inside wall of the pipe (see *Figure 45*). This can cause additional turbulence in flow once the system is placed in operation, and must be considered during the design stages when many such joints exist.

Follow the manufacturer's suggested procedures on the proper setup and operation of the roll-grooving tool.

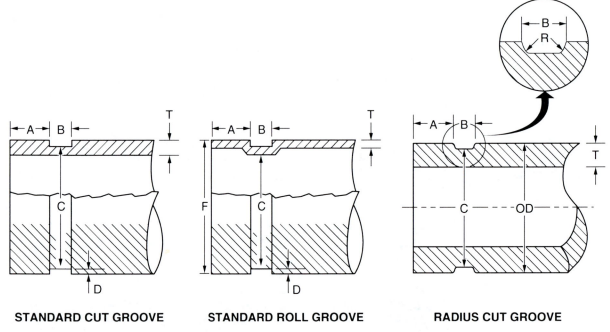

Figure 45 ◆ Details of pipe grooves.

USED WITH PERMISSION OF VICTAULIC COMPANY OF AMERICA.
406F46.EPS

Figure 46 ♦ Power roll-grooving machine.

7.1.2 Cut Grooving

Cut grooving differs from roll grooving in that a groove is cut into the pipe. Unlike roll grooving, material is removed from the pipe, and the wall thickness at the groove becomes thinner. However, the inside wall of the pipe has no deformity to create turbulence. Cut grooving is intended for standard weight or heavier pipe. The cut removes less than one-half the pipe wall, which is less depth than thread cuts. Cut-grooving machines are designed to be driven around a stationary pipe. This ensures a groove that is concentric with the pipe OD and is of uniform depth. The cut-grooving tool is driven by an external power source.

Follow the manufacturer's suggested procedures on the proper setup and operation of the cut grooving tool.

7.2.0 Selecting Gaskets

There are many types of synthetic rubber gaskets available to provide the option of selecting grooved piping products for the widest range of applications. In order to provide maximum life for the service intended, the proper gasket selection and specification is required.

Several factors must be considered in determining the best gasket to use for a specific service. The foremost consideration is the temperature of the product flowing through the pipe. Temperatures beyond the recommended limits decrease gasket life. Also, the concentration of the product, duration of service, and continuity of service must be considered because there is a direct relationship between temperature, continuity of service, and gasket life.

It should also be noted that there are services for which the gaskets are not recommended. Always check the latest gasket selection guide for specific gasket service recommendations and for a listing of services that are not recommended. *Table 5* shows standard and special gasket applications.

> **NOTE**
> Gasket selection recommendations apply only to Victaulic coupling gaskets. These recommendations do not necessarily apply to Victaulic rubber-lined valves or other rubber-lined products. Consult Victaulic for recommendations.

7.3.0 Installing Grooved Pipe Couplings

Follow these steps to install a typical grooved pipe coupling.

Step 1 Ensure that you are using the proper size and type of gasket.

Step 2 Apply a thin coat of lubricant or silicone to the gasket lips and to the outside of the gasket.

Step 3 Place the gasket over the pipe end.

Step 4 Arrange the gasket so that it does not hang over the pipe end.

Step 5 Align and bring the two pipe ends together.

Step 6 Center the gasket between the grooves on each pipe.

Step 7 Ensure that no portion of the gasket is allowed to extend into the groove on either pipe.

Step 8 Place housings over the gasket.

Step 9 Engage the housing keys into the grooves.

Step 10 Insert the bolts, and tighten the nuts finger-tight.

Step 11 Tighten the nuts alternately and equally until the housing bolt pads are firmly together.

> **CAUTION**
> Excessive nut tightening is not necessary.
> Uneven tightening may cause the gasket to pinch.

Table 5 Standard and Special Gasket Applications

Grade	Temp. Range	Compound	Code	*General Service Recommendations
Standard Gaskets IPS				
E	−30°F to +230°F −34°C to +110°C	EPDM	Green Stripe	Recommended for hot water service within the specified temperature range plus a variety of dilute acids, oil-free air, and many chemical services. UL classified in accordance with ANSI/NSF 61 for cold +86°F/+30°C and hot +180°F/+82°C potable water service. **Not recommended for petroleum services.**
T	−20°F to +180°F −29°C to +82°C	Nitrile	Orange Stripe	Recommended for petroleum products, hydrocarbons, air with oil vapors, vegetable and mineral oils within the specified temperature range; not recommended for hot dry air over +140°F/+60°C and water over +150°F/+66°C. **Not recommended for water services.**
E[†] (Type A)	Ambient	EPDM	Violet Stripe	Applicable for wet and dry (oil-free air) sprinkler services only. For dry services, Victaulic continues to recommend the use of FlushSeal® gaskets. **Not recommended for hot water services.**
Special Gaskets IPS				
M2	−40°F to +160°F −40°C to +71°C	Epichlorohydrin	White Stripe	Specially compounded to provide superior service for common aromatic fuels at low temperatures. Also suitable for certain ambient temperature water services.
V	−30°F to +180°F −34°C to +82°C	Neoprene	Yellow Stripe	Recommended for hot lubricating oils and certain chemicals. Good oxidation resistance. Will not support combustion.
O	+20°F to +300°F −7°C to +149°C	Fluoro-elastomer	Blue Stripe	Recommended for many oxidizing acids, petroleum oils, halogenated hydrocarbons, lubricants, hydraulic fluids, organic liquids and air with hydrocarbons to +300°F/+149°C. **Not recommended for hot water services.**
L	−30°F to +350°F −34°C to +177°C	Silicone	Red Gasket	Recommended for dry heat, air without hydrocarbons to +350°F/+177°C, and certain chemical services.
A	+20°F to +180°F −7°C to +82°C	White Nitrile	White Gasket	No carbon black content. May be used for food. Meets FDA requirements. Conforms to CFR Title 21 Part 177.2600. Not recommended for hot water services over +150°F/+66°C or for hot, dry air over +140°F/+60°C. **Not recommended for hot water services.**
T EndSeal	−20°F to +150°F −29°C to +66°C	Nitrile	No External Identification	Specially compounded with excellent oil resistance and a high modulus for resistance to extrusion. Temperature Range −20°F/−29°C to +150°F/+66°C. Recommended for petroleum products, air with oil vapors, vegetable and mineral oils within the specified temperature range. **Not recommended for hot water services over +150°F/+66°C or for hot, dry air over +140°F/+60°C.** For maximum gasket life under pressure extremes, temperature should be limited to +120°F/+49°C.
EG	−30°F to +230°F −34°C to +110°C	EPDM	Double Green Stripes	Recommended for hot water service withing the specified temperature range plus a variety of dilute acids, oil-free air, and many chemical services. DVGW, KTW, OVGW, and SVGW approved for W534, EN681-1 Type WA cold potable water service up to +122°F/+50°C. **Not recommended for petroleum services.**
EF	−30°F to +104°F −34°C to +40°C	EPDM	Green X	Recommended for potable water service within the specified temperature range plus a variety of dilute acids, oil-free air and many chemical services. French ACS (Crecep) approved for EN681-1 Type WA cold potable water service. Not recommended for petroleum services.
Special Gaskets AWWA and Transition Coupling				
S	−20°F to +180°F −29°C to +82°C	Nitrile	Red Stripe	Specially compounded to conform to ductile pipe surfaces. Recommended for petroleum products, air with oil vapors, vegetable and mineral oils within the specified temperature range; not recommended for hot dry air over +140°F/+60°C and water over +150°F/+66°C. **Not recommended for hot water services.**
M	−20°F to +200°F −29°C to +93°C	Halogenated Butyl	Brown Stripe	Recommended for water service within the specified temperature range plus a variety of dilute acids, oil-free air and many chemical services. Readily conforms to ductile pipe surfaces. UL classified in accordance with ANSI/NSF 61 for cold +86°F/+30°C potable water service. **Not recommended for petroleum services.**

† Vic-Plus gasket. Lubrication not required
* For specific chemical and temperature compatibility, refer to the Gasket Selection and Chemical Services sections.
The information shown defines general ranges for all compatible fluids.

Review Questions

1. Copper tubing can be connected by _____.
   a. flared or compression joints
   b. rod welding
   c. stick welding
   d. oxyacetylene welding

2. The color code for Type L copper tubing is _____.
   a. red
   b. gold
   c. blue
   d. green

3. One reason for using flared joints rather than soldered joints is they _____.
   a. are cheaper
   b. can be disassembled and reassembled easily
   c. are stronger
   d. are faster to put together

4. The difference between Type K and Type L copper tubing is _____.
   a. Type K has thinner walls
   b. Type L has thinner walls
   c. Type K is drawn tubing
   d. Type L is a different metal

5. Long flare nuts are preferred over short flare nuts for _____.
   a. potable water
   b. high vibration applications
   c. carbon steel tubing
   d. aluminum tubing

6. The ring that is compressed in a compression fitting is called a _____.
   a. nut
   b. flange
   c. ferrule
   d. washer

7. The main difference between soldering and brazing is the _____.
   a. type of tubing
   b. temperature of the heating procedure
   c. type of filler used
   d. depth of the cup

8. When assembling long runs of tubing, the tubing must be _____.
   a. properly supported
   b. made out of shorter pieces
   c. no longer than 20 feet
   d. allowed to hang loose

9. Soldered joints depend on _____.
   a. heat above 800 degrees
   b. capillary action
   c. quenching
   d. shallow cups

10. Soldering flux cleans and protects the tubing from _____.
    a. penetration
    b. oxidation
    c. adhesion
    d. osmosis

11. Wetting is the process that decreases the _____ so that the solder flows into the joint.
    a. penetration
    b. gravitation
    c. surface tension
    d. oxidation

12. The best fluxes for joining copper tubing and copper fittings are _____.
    a. corrosive fluxes
    b. inflammable
    c. lead based
    d. non-corrosive

13. The BAg series of filler metals contain _____.
    a. silver
    b. phosphorus
    c. lead
    d. aluminum

14. BAg-1 and BAg-2 fillers also contain _____, and can be hazardous to breathe when heated.
    a. mercury
    b. cadmium
    c. oxygen
    d. iron

Review Questions

15. To calculate the tubing needed for a brazed or soldered pipe, add the _____ to the face-to-face measurement.
    a. tangent
    b. cup depth engagement
    c. solder thickness
    d. length of the fitting

16. Bending pipe allows the process to _____.
    a. cost less
    b. heat up more
    c. be stopped
    d. flow more freely

17. The formula used to determine the bend allowance is _____.
    a. L = R − D*0.01745
    b. L = R*D*0.01745
    c. L = R*sinD
    d. L = R*tanD

18. The bend allowance for a 45-degree bend on a 24-inch radius is _____.
    a. 18⅞ inches
    b. 24⅞ inches
    c. 36½ inches
    d. 48³⁄₁₆ inches

19. The point where two straight lines meet to form an angle is the _____.
    a. tangent
    b. arc
    c. vortex
    d. vertex

20. The inside radius of a pipe bend is calculated by subtracting _____ from the center line radius.
    a. half the OD
    b. twice the OD
    c. the sine of the degree of bend
    d. the OD

21. Glass-lined pipe is used in _____.
    a. foundries
    b. chemical and food processing industries
    c. power plants
    d. mining

22. When working with glass-lined pipe, *never* _____.
    a. use a torque wrench
    b. weld to the outside
    c. use a gasket
    d. empty the pipe

23. When preparing a Lokring® joint, sand _____.
    a. the inside of the fitting
    b. around the circumference of the pipe
    c. down the axis of the pipe
    d. the edge of the pipe

24. The larger end of the Lokring® NO GO gauge is used to _____.
    a. test for correct diameter
    b. mark the pipe
    c. compress the tubing
    d. flare the end

25. Flexible grooved couplings allow for _____.
    a. expansion
    b. reduction
    c. loss of support
    d. loose bolts

Summary

Throughout your career as a pipefitter, you will work with various types of specialized piping materials and processes. This module has introduced you to some of the more common types of special piping, but it would be impossible to cover the constantly changing technology of processes and materials in the field. Before attempting to install any type of special piping, always refer to the manufacturer's installation and maintenance guidelines.

Notes

Trade Terms Introduced in This Module

Acetylene: A gas composed of two parts carbon and two parts hydrogen, commonly used in combination with oxygen to cut, weld, and braze steel.

Compression coupling: A mechanical fitting that is compressed onto a pipe, tube, or hose.

Ferrule: A bushing placed over a tube to tighten it.

Flaring: Increasing the diameter at the end of a pipe or tube.

Nonferrous metal: A metal that contains no iron and is therefore nonmagnetic.

Oxidation: A chemical reaction that increases the oxygen content of a compound.

Reamer: A tool used to enlarge, shape, smooth, or otherwise finish a hole.

Solder: An alloy, such as of zinc and copper, or of tin and lead, that, when melted, is used to join metallic surfaces.

Soldering flux: A chemical substance that aids the flow of solder. Flux removes and prevents the formation of oxides on the pieces to be joined by soldering.

Wetting: Spreading liquid filler metal or flux on a solid base metal.

Resources & Acknowledgments

Additional Resources

This module is intended to present thorough resources for task training. The following reference works are suggested for further study. These are optional materials for continued education rather than for task training.

Victaulic Field Assembly and Installation Instruction Pocket Handbook I-100, Victaulic Company of America, P.O. Box 31, Easton, PA 18044-0031, (610) 559-3300.

LP-101 or LP-105 Installation Procedure for Lokring Type 316/316 L Fittings, Lokring Corporation, 396 Hatch Drive, Foster City, CA 94404, (415) 578-9999.

Figure Credits

Topaz Publications, Inc., 406F02, 406F04 (photo), 406F05, 406F08 (photo), 406F10, 406F11 (photo)

Van Sant Enterprises/TrickTools.com, 406F26

LOKRING Technology Corporation, 406F32-406F41

Victaulic Companies, 406F42, 406F45, 406F46, 406T05

Ridge Tool Co. (Ridgid®), 406F43, 406F44

NCCER CURRICULA — USER UPDATE

NCCER makes every effort to keep its textbooks up-to-date and free of technical errors. We appreciate your help in this process. If you find an error, a typographical mistake, or an inaccuracy in NCCER's curricula, please fill out this form (or a photocopy), or complete the online form at **www.nccer.org/olf**. Be sure to include the exact module ID number, page number, a detailed description, and your recommended correction. Your input will be brought to the attention of the Authoring Team. Thank you for your assistance.

Instructors – If you have an idea for improving this textbook, or have found that additional materials were necessary to teach this module effectively, please let us know so that we may present your suggestions to the Authoring Team.

NCCER Product Development and Revision
13614 Progress Blvd., Alachua, FL 32615

Email: curriculum@nccer.org
Online: www.nccer.org/olf

❏ Trainee Guide   ❏ AIG   ❏ Exam   ❏ PowerPoints   Other _____

Craft / Level: Copyright Date:

Module ID Number / Title:

Section Number(s):

Description:
_____
_____
_____
_____

Recommended Correction:
_____
_____
_____
_____

Your Name:

Address:

Email: Phone:

Pipefitting Level Four

08407-07

Hot Taps

08407-07
Hot Taps

Topics to be presented in this module include:

1.0.0	Introduction	7.2
2.0.0	Hot Tap Safety and Potential Hazards	7.2
3.0.0	Installing Fittings	7.2
4.0.0	Operating Hot Tap Machines	7.5
5.0.0	Line Stop Plugs	7.8

Overview

This module describes the methods and technologies for connecting to pipe that is not to be shut off or emptied. In such a circumstance, the pressure of the fluid must be contained, to prevent leaks. Different contents require different methods of hot tapping. The module describes methods of penetrating pipes that transport liquids while the connections are made. There are also sections on methods of temporarily stopping flow while connections are made, such as line stop plugs, pipe freezing, and pipe plugging.

Objectives

When you have completed this module, you will be able to do the following:

1. Explain hot tap safety and potential hazards.
2. Identify fittings used with hot taps.
3. Explain the use of hot tap machines.
4. Identify and explain the use of stopples.

Trade Terms

Full port
Head clearance
Hot tap
Line stop
Travel distance

Required Trainee Materials

1. Pencil and paper
2. Appropriate personal protective equipment

Prerequisites

Before you begin this module, it is recommended that you successfully complete *Core Curriculum; Pipefitting Level One; Pipefitting Level Two; Pipefitting Level Three;* and *Pipefitting Level Four*, Modules 08401-07 through 08406-07.

This course map shows all of the modules in the fourth level of the Pipefitting curriculum. The suggested training order begins at the bottom and proceeds up. Skill levels increase as you advance on the course map. The local Training Program Sponsor may adjust the training order.

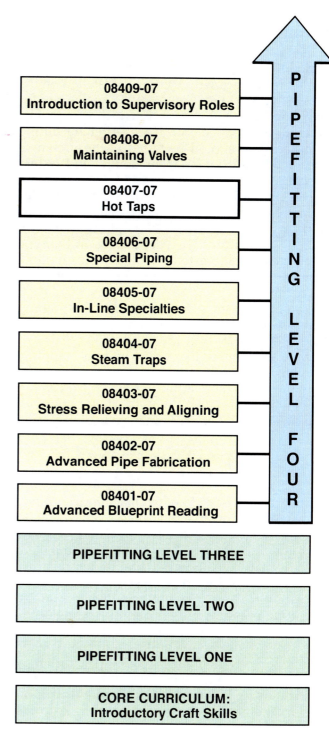

1.0.0 ♦ INTRODUCTION

Hot tap machines are drilling and tapping machines used when a branch connection must be added to an existing pipeline, tank, or vessel that cannot be shut down. A fitting is installed on the pipe or vessel to secure the tap. Usually, the fitting is welded, but in cases in which it cannot be welded, it is bolted into place. After the fitting is installed, it should be tested at the pressures specified by the hot tap manufacturer and system specifications.

Hot tap machines can be used on different types of metals, such as carbon steel, some alloys, cast iron, or ductile iron, as well as plastics. The hot tapping procedures vary depending on the type of pipe the hot tap is to be made on. This module explains how to install fittings and line stops and make hot taps.

You may never perform a hot tap. The training that is required for a particular machine, as well as the expense of the machine itself, makes specialty contractors more likely than in-house tapping. In addition, insurance considerations may make the contractors more attractive still. However, hot tapping is a specialty within the pipefitting field, and this module serves as an introduction.

2.0.0 ♦ HOT TAP SAFETY AND POTENTIAL HAZARDS

There are a variety of hot tap machines, and they all have common characteristics. Each is able to contain the pressure inside an existing pipeline until the branch connection has been completed. Each hot tap machine has a specific temperature and pressure rating that must not be exceeded for safety reasons. These ratings depend on the operating pressure and temperature rating of the pipeline and the branch being installed.

A drilling device for drilling into the pipeline is common on all hot tap machines. The drill may use flat drills, high-speed drills, or hole saws. Each hole saw is equipped with a pilot drill or bit to start the hole in the pipeline and to keep the cutter square during the operation.

When selecting a hot tap machine to perform a hot tap, consider the following:

- Size of the hot tap to be made
- Pressure rating of the valve flange
- Thickness of the pipe or vessel
- Size of the pipe or vessel
- Material the pipe or vessel is made of
- Pressure inside the pipe or vessel
- Temperature inside the pipe or vessel
- Product inside the pipe or vessel
- Angle of the tap
- Head clearance for the hot tap machine
- Travel distance of the drill bit

The requirement for head clearance for a hot tap on pipe as large as ten inches in diameter may be fourteen feet or more. The drill bit must travel through the valve and through the wall of the pipe, far enough, in fact, to take the hole saw all the way through the wall of the pipe, without cutting the far wall.

> **WARNING!**
> Improper use of hot tap machines can cause serious personal injury.

For safe use of hot tap machines, follow the basic rules listed below:

- Choose the proper hot tap machine for the job.
- Never use a tool for anything other than its intended use.
- Wear eye protection, either safety glasses or goggles.
- Wear a welding shield during welding operations.
- Wear gloves when using tools that require them.
- Observe welding codes whenever welding is being done.
- Never weld a fitting on a tank in an area where vapors may exist.
- Determine the material flowing through a pipeline before servicing the line.
- Barricade areas according to company policies.
- Follow the manufacturer's instructions. You may have to receive special training on a particular process or type of equipment.

3.0.0 ♦ INSTALLING FITTINGS

A fitting must be installed on the existing pipeline before a hot tap can be made. It is important that the fitting be installed properly and that welding codes be observed. Fittings are used where single branch pipelines are to be tapped. Tapping valves are installed between the hot tap machine and the fitting on the pipe. The valve must be a full port gate valve so that the hot tap cutter will clear all sides of the valve when installed. After installing the valve to the fitting, the valve must be pressure-tested according to plant specifications. There are three basic types of fittings used with hot tap machines: mechanical joint, bolt-weld, and split tee fittings.

3.1.0 Installing Mechanical Joint Fittings

Mechanical joint fittings bodies are supports for the piping and come with six pieces, forming two halves that are placed around the pipeline where the hot tap is to be made. The mechanical joint fitting has the two halves of the body, which are to be bolted together on the pipe, and the glands that complete the fitting are also made in two halves that bolt together. *Figure 1* shows a mechanical joint fitting. There are specific temperature limitations for mechanical joint fittings, due to the material of the gasket.

Follow these steps to install a mechanical joint fitting:

Step 1 Clean the pipeline to remove any excess dirt, using a nonflammable solvent.

Step 2 Clean the mechanical joint fitting to remove any excess dirt, using a nonflammable solvent. Ensure that the pipe is not egg-shaped in case one uses a stopper.

Step 3 Place the top half of the mechanical joint fitting on top of the pipeline where the hot tap is to be drilled.

Step 4 Place the bottom half of the mechanical joint fitting under the pipeline.

Step 5 Fasten the top and bottom halves of the mechanical joint fittings together, using the bolts that are provided with the fitting, leaving about ¼ inch between the top and bottom mechanical joint flanges.

Step 6 Slip an oval, rubber-side gasket into the groove between the flanges so that the gasket extends about 1 inch beyond the end of the mechanical joint fitting.

Step 7 Tighten the flange bolts, using a socket wrench, until the flanges are tight.

Step 8 Cut off the ends of the gasket using a knife, leaving about ⅛ inch of the gasket protruding into the socket.

Step 9 Center the mechanical joint fitting on the pipe, with the outlet in the proper position.

Step 10 Place end gaskets around the pipe, with the tapered end of the gaskets facing into the socket.

Step 11 Press the end gaskets into position.

Step 12 Assemble the glands on the pipe.

Step 13 Insert the bolts.

Step 14 Center the glands against the end gaskets.

Step 15 Tighten the flange bolts evenly to compress the end gaskets against the pipe, using a torque wrench.

WARNING!
Follow the appropriate torquing procedures for the type of mechanical joint being installed. Failure to torque the bolts properly could result in serious injury when pressure is applied to the pipe and mechanical joint fitting.

Step 16 Install the appropriate valve on the mechanical joint fitting.

Step 17 Conduct the appropriate pressure test for the mechanical joint fitting and valve assembly.

3.2.0 Installing Bolt-Weld Fittings

Bolt-weld fittings have three major parts: the seal, the top half of the fitting, and the bottom half of the fitting. Bolt-weld fittings are used in a pipeline to facilitate repairs, modifications, or additions and are also used with hot tap machines to make branch connections. They can be used with stopples with or without a temporary bypass. *Figure 2* shows the assembly process for a bolt-weld fitting.

Follow these steps to install a bolt-weld fitting:

Step 1 Clean the pipeline to remove any excess dirt using a nonflammable solvent.

Step 2 Clean the bolt-weld fitting to remove any excess dirt using a nonflammable solvent.

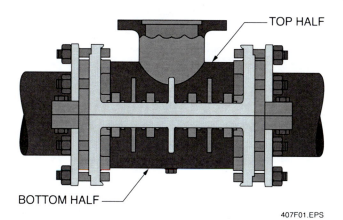

Figure 1 ♦ Mechanical joint fitting.

Step 3   Place the seal on the pipeline.

Step 4   Place the top half of the fitting on top of the seal.

Step 5   Place the bottom half of the fitting under the pipeline.

Step 6   Tighten the socket head cap screws to hold the bottom half in place.

Step 7   Fillet-weld the top and bottom halves together, and to the pipe if specified.

> **WARNING!**
> Follow the proper welding procedures and codes for the given pipeline to avoid personal injury.

3.3.0 Installing Split Tee Fittings

A split tee fitting (*Figure 3*) is a mechanical device used to hold the tapping valve and machine during the tapping procedure. Split tee fittings are welded to the pipe. Split tee fittings are installed to aid in branch line taps, plugging taps, and new construction where branch connections require reinforcement around the entire circumference of the pipe.

Follow these steps to install a split tee fitting:

Step 1   Ensure that each piece of the split tee has a tapped hole. This is to provide a port to insert a nipple into to pressure-test each weld when the tee fitting is installed.

Step 2   Ensure that the pipe has been ultrasound tested for thickness.

Step 3   Ensure that there is flow through the line before welding.

> **WARNING!**
> Never weld onto a stagnant pipeline. The flow through the line is needed to displace the heat from welding.

Step 4   Determine the temperature, pressure, and thickness of the pipe.

Step 5   Obtain all necessary permits.

Step 6   Place the split tee fitting on the pipeline where the connection is to be made.

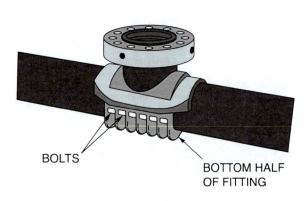
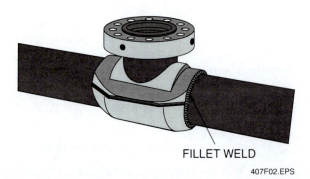

Figure 2 ◆ Assembly process for a bolt-weld fitting.

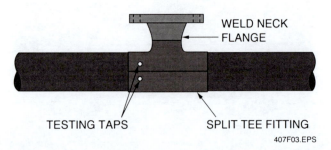

Figure 3 ◆ Split tee fitting.

Step 7 Tack-weld the split tee fitting along its seams to the pipeline.

CAUTION
Follow the proper welding procedures and codes for the type of material being used. Protect the threads of the tapped hole when welding.

Step 8 Grind the tacks smooth, using a 14-inch half-round bastard file.

CAUTION
Be careful not to grind into the base metal of the pipe.

Step 9 Fillet-weld the split tee fitting along its seams to the pipeline.

WARNING!
Stagger the welding process from one part of the pipe to the other so you will not have an excessive amount of heat in one area.

Step 10 Test the split tee through the tapped holes, using the proper procedures.

Step 11 Determine if you are going to seal-weld or insert screw plugs into the testing taps.

Step 12 Seal the testing taps.

4.0.0 ♦ OPERATING HOT TAP MACHINES

All hot tap machines consist of the same basic components: a drive mechanism that is either pneumatically, hydraulically, or manually powered and that is used to drive the hole saw; a drive shaft that connects the drive mechanism to the cutter; a machine adapter that is used to attach the hot tap machine to the tapping valve; a cutter that is used to cut a hole in the pipe wall and remove a piece of pipe known as the coupon from the pipe wall.

When performing a hot tap, you must always remove and save the pipe coupon so that it can be inspected by the safety engineer. Do not allow the coupon to be lost inside the pipeline. Hole saws are designed with a U-shaped wire that is retracted in the saw until the saw cuts through the pipe wall. After passing through the pipe wall the U-shaped wire protrudes from the hole saw to hold the coupon so that it can be removed.

Before operating a hot tap machine, you must receive training on the specific machine that you are using. The two basic types of hot tap machines are hand-operated and power-operated.

4.1.0 Operating Hand-Operated Hot Tap Machines

Hand-operated hot tap machines are normally used in areas that are highly explosive because you can control the speed of the tap. They are operated with a ratchet mechanism or a wrench to advance the drill or cutter. A measuring rod is used to ensure that the proper travel distance of the cutter is achieved when it is lowered. *Figure 4* shows a hand-operated hot tap machine.

Follow these steps to operate a hand-operated hot tap machine:

WARNING!
This procedure is provided to teach you how hot tap machines operate. Never attempt to operate a hot tap machine unless you have received specialized training from the equipment manufacturer.

Step 1 Inspect the valve and fitting to be sure no foreign material is present.

Step 2 Check the valve to be sure it opens freely.

Step 3 Check the hot tap machine manufacturer's chart to determine the proper machine travel.

Step 4 Attach the appropriate tool to the boring bar.

Step 5 Bolt the flanges of the hot tap machine and the valve to connect the hot tap machine and the valve.

Step 6 Turn the ratchet clockwise to lower the boring bar and make the hot tap.

Step 7 Turn the ratchet counterclockwise to raise the boring bar.

Step 8 Close the valve.

Step 9 Remove the bolts from the hot tap machine and valve flanges to disconnect the hot tap machine from the valve.

4.2.0 Operating Power-Operated Hot Tap Machines

Power-operated hot tap machines function like the hand-operated hot taps except that the drill is

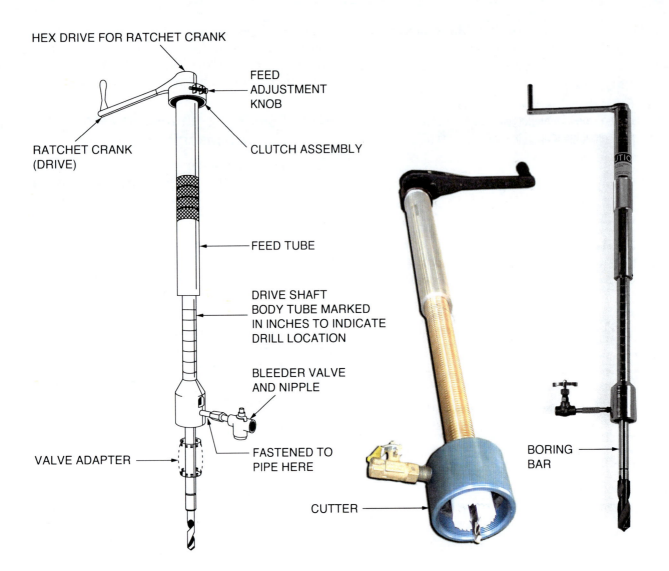

NOTE: Valve adapters are used to adapt machine to various size valves.

Figure 4 ♦ Hand-operated hot tap machine.

hydraulically or pneumatically powered. *Figure 5* shows a power-operated hot tap machine.

Follow these steps to operate a power-operated hot tap machine:

 WARNING!
This procedure is provided to teach you how hot tap machines operate. Never attempt to operate a hot tap machine unless you have received specialized training from the equipment manufacturer.

Step 1 Bolt the flanges of the hot tap machine and valve to connect the hot tap machine and the valve.

Step 2 Activate the appropriate hot tap machine switch to lower the boring bar into place.

Step 3 Check the alignment of the boring bar in relation to the pipeline and adjust the boring bar as needed.

Step 4 Activate the appropriate hot tap machine switch to raise the boring bar.

Step 5 Attach the appropriate cutter/reamer tool to the boring bar.

Step 6 Activate the appropriate hot tap machine switch to lower the boring bar and begin the cutting/reaming process.

Step 7 Activate the appropriate hot tap machine switch to raise the cutter/reamer tool.

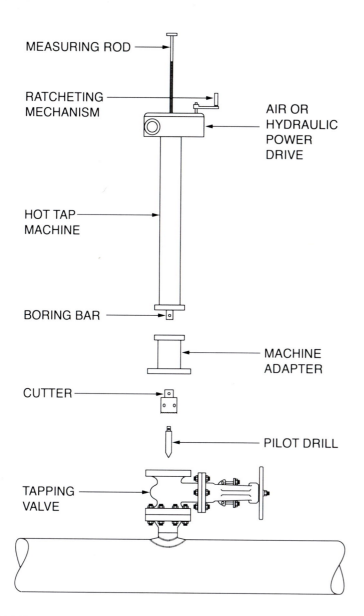

Figure 5 ♦ Power-operated hot tap machine.

Step 8 Close the valve.

Step 9 Release pressure from the fluid trapped between the valve and the hot tap machine.

Step 10 Drain the contents of the valve according to plant procedures and OSHA regulations.

Step 11 Remove the bolts from the hot tap machine and valve flanges to disconnect the hot tap machine from the valve.

Step 12 Blind-flange the valve unless a spool is to be added to this point.

4.3.0 New Technologies in Hot Tapping

The spread of plastic pipe in many uses, including gas transport, has led to the development of equipment specifically developed for hot tapping plastic. Some such equipment allows the branch to be fusion welded to the main pipe as part of the process, speeding the hot tap procedure.

Another development is that of diverless systems for hot tapping undersea pipe. The system in use employs cradles to lift the pipe clear of the bottom, to keep out the muck and sand on the bottom. When the pipe is in place, a remote-controlled hot tap machine is deployed, allowing the pipe to be entered at depths not easily accessed by divers.

Another place where hot tapping equipment has found a use is in draining sunken tanker vessels underwater. A hot tap fitting and valve are attached to the hull and the hole is drilled, then the valve is shut off. The machine is removed, a hose is attached, and the stored fluids pumped to tanker vessels on the surface

5.0.0 ◆ LINE STOP PLUGS

Line stop plugs (*Figure 6*) serve as temporary block valves and can be installed anywhere in the system. They are used to isolate a section of line for repairs or additions without interrupting service. T.D. Williamson, Inc., probably the largest company in the field, calls this equipment STOPPLE® fittings, and that has become the slang term for all such devices. The installation and use of line stop plugs is a highly specialized skill and should only be done by trained, certified personnel. Before installing the plug, you must know the following information:

- The size of the hot tap
- The size of the valve bore
- The pressure rating of the valve
- Whether the valve is flanged or screw-type
- Whether the valve is operating properly
- The distance from the outside face of the valve to the pipe
- The clearance on the machine
- What material the pipe is made of
- The thickness of the pipe, vessel, or tank to be tapped so that you can determine if it is safe to weld or bolt a fitting to the pipe without crushing it or burning a hole in it
- The pressure inside the pipe, vessel, or tank to be tapped
- The temperature of the product inside the pipe, vessel, or tank to be tapped
- The rigging needs

Line stop plugs are used on low-, medium-, and high-pressure piping systems as well as on vacuum systems. Line stops use a variety of sealing elements that are selected based on compatibility with the piping service. Line stops are used after a hot tap has been completed and the coupon has been removed. After the line stop is completed, a completion plug is installed and is locked and held in place through the use of jack screws in the custom-designed split tee fitting so that the tapping

Courtesy of TD Williamson, Inc.
407F06.EPS

Figure 6 ◆ Line stop plugs.

valve can be removed. Line stops can be installed in pipes ranging in size from ½ to 36 inches in diameter and enable pipelines to be isolated for repairs or additions without interrupting service or losing product. *Figure 7* shows a plugging machine.

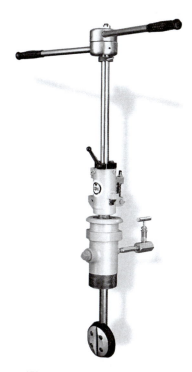

Figure 7 ◆ Line stop plugging machine.

5.1.0 Freeze Stops

Freeze stops (*Figure 8*), or cryogenic plugs, are used to temporarily plug lines containing freezable liquids under pressure. Freeze stops differ from other stops in that they use a chemical process rather than a mechanical process and therefore do not require the use of a hot tap machine. A freeze stop is the controlled formation of a solid plug inside a pipe containing flowing liquids. The pipeline is frozen so that the pipe diameter shrinks slightly and liquid crystals adhere to the inside wall, thus plugging the pipeline. This shrinkage and adherence enables the plug to maintain its position and resist high pressures that may exist in the pipeline.

Follow these steps to assist a technician to perform a freeze stop:

Step 1 Tag out the area around the pipeline to be isolated in accordance with the local procedures.

Step 2 Ensure that the temperature-monitoring equipment is in proper working condition.

Step 3 Ensure that the pipeline pressure is low as possible.

Step 4 Ensure that the pipeline temperature is low as possible.

Step 5 Ensure that the pipe section where the freeze stop is to be performed is full of a freezable liquid that can be stopped from flowing.

Step 6 Clean the pipe where the freeze stop is to be performed, being sure to remove any paint from the area.

Step 7 Assist the technical representative to perform the freeze stop.

5.2.0 Cross Line Stops

Cross line stops are often used when there is not enough clearance for a conventional line stop of a branch line coming off a main line or header. Installing a cross line stop involves making a hot tap from the back side of the pipeline opposite the branch line. The branch is then plugged from the inside of the main line or header. A guide attached to the tap machine boring bar is lowered into place to provide the seal. This type of stopple can withstand temperatures up to 1,200°F and pressures up to 6,000 psi. *Figure 9* shows a typical cross line stop.

5.3.0 Low-Pressure and Vacuum Stops

Plugging low-pressure (below 100 psi) or negative pressure (vacuum) piping systems requires that four split tee fittings and two equalization fittings be installed. Two split tee fittings are installed as line stops and two are installed as bypass fittings. *Figure 10* shows a typical setup for low-pressure and vacuum stops.

Figure 8 ◆ Freeze stop.

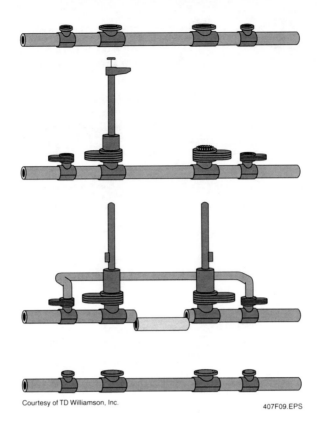

Figure 9 ◆ Cross line stop.

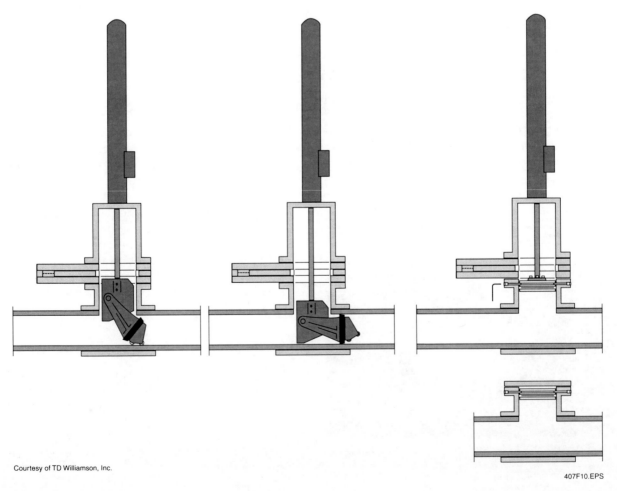

Figure 10 ◆ Setup for low-pressure and vacuum stops.

7.10 PIPEFITTING ◆ LEVEL FOUR

Review Questions

1. Each hot tap machine is able to _____.
   a. contain the pressure in a pipeline
   b. operate at any pressure
   c. operate without an attendant
   d. work on any kind of pipe

2. The valve used in a hot tap must be a _____ valve.
   a. butterfly
   b. plug
   c. check
   d. full port gate

3. The mechanical joint hot tap fitting has _____.
   a. no limitations
   b. specific temperature limitations
   c. only one piece
   d. one piece glands

4. The three major parts of a bolt-weld fitting are _____.
   a. glands, gasket, and body
   b. drill, cap, and valve
   c. seal, top half, and bottom half
   d. flange, gasket, and seal

5. The testing taps on a split tee are used to _____.
   a. insert a thermowell
   b. drill holes in the pipe
   c. test the welds
   d. look at the pipe

6. The flow through a pipeline is needed to _____.
   a. carry away the coupons
   b. displace the heat from the welding
   c. keep the drill lubricated
   d. heat the pipe

7. The basic components of a hot tap machine are _____.
   a. a flange, a gasket and a valve
   b. the drive mechanism, shaft, adapter, and cutter
   c. handle, grinder and flange
   d. the drill, a valve and a gasket

8. When you perform a hot tap, you must save the _____.
   a. shavings
   b. material from the bleed valve
   c. pipe coupon
   d. seal

9. To make the tap, hand-operated hot tap machines are turned _____.
   a. upward
   b. clockwise
   c. downward
   d. counterclockwise

10. Line stop plugs are used to _____.
    a. isolate a section of line
    b. install a branch line
    c. increase line pressure
    d. increase flow rate

Summary

Pipefitters are often called upon to add a branch connection or do other maintenance type work to an existing pipeline, tank, or vessel that cannot be shut down. In order to do this, the pipefitter must know how to assist a trained technician to perform a hot tap, including knowing what type of information that the manufacturer needs to recommend the correct hot tap machine and/or line stop plug. Normally, hot tap and line stop plug manufacturers can provide you with a checklist to fill out that will provide them with the information they need. The pipefitter must also know the potential hazards and safety guidelines to follow when performing a hot tap or a freeze stop.

Notes

Trade Terms Introduced in This Module

Full port: The maximum internal opening for flow through a valve that matches the ID of the pipe used.

Head clearance: The amount of space needed to install a hot tap machine on a pipe.

Hot tap: To make a safe entry into a pipe or vessel operating at a pressure or vacuum under controlled conditions without losing product.

Line stop: A device used to temporarily contain the flow or pressure of a product inside a pipe while a branch connection is being made.

Travel distance: The distance that the cutting bit moves from the top of the tapping valve to the pipe to be cut.

Resources & Acknowledgments

Additional Resources

This module is intended to present thorough resources for task training. The following reference work is suggested for further study. This is optional material for continued education rather than for task training.

www.midwestpiperepair.com/page2.html.

Figure Credits

IFT Products, **www.tappingmachines.com**, 407F04 (middle), 407F05 (photo)

T.D. Williamson, Inc., 407F04 (right), 407F06–407F10

NCCER CURRICULA — USER UPDATE

NCCER makes every effort to keep its textbooks up-to-date and free of technical errors. We appreciate your help in this process. If you find an error, a typographical mistake, or an inaccuracy in NCCER's curricula, please fill out this form (or a photocopy), or complete the online form at **www.nccer.org/olf**. Be sure to include the exact module ID number, page number, a detailed description, and your recommended correction. Your input will be brought to the attention of the Authoring Team. Thank you for your assistance.

Instructors – If you have an idea for improving this textbook, or have found that additional materials were necessary to teach this module effectively, please let us know so that we may present your suggestions to the Authoring Team.

NCCER Product Development and Revision
13614 Progress Blvd., Alachua, FL 32615

Email: curriculum@nccer.org
Online: www.nccer.org/olf

❑ Trainee Guide ❑ Lesson Plans ❑ Exam ❑ PowerPoints Other _____

Craft / Level: _____ Copyright Date: _____

Module ID Number / Title: _____

Section Number(s): _____

Description: _____

_____

_____

_____

_____

Recommended Correction: _____

_____

_____

_____

_____

Your Name: _____

Address: _____

_____

Email: _____ Phone: _____

Pipefitting Level Four

08408-07

Maintaining Valves

08408-07
Maintaining Valves

Topics to be presented in this module include:

1.0.0 Introduction 8.2
2.0.0 Removing and Installing Valves 8.2
3.0.0 Valve Stems and O-Rings 8.8
4.0.0 Replacing Bonnet Gaskets 8.11
5.0.0 Packing Valves 8.13

Overview

This module describes the methods used to maintain valves. Here are the methods for the safe assembly and repair of valves, as well as troubleshooting procedures. The procedure for replacing packing and O-rings is covered, as well as the types of bonnets and stems. The assembly and disassembly of several different varieties of valve are taught, with descriptions of various problems and hazards specific to each.

Objectives

When you have completed this module, you will be able to do the following:

1. Remove and install threaded valves.
2. Remove and install flanged valves.
3. Replace valve stem O-rings.
4. Replace bonnet gaskets.
5. Explain the purpose of valve packing.
6. Explain or demonstrate how to repack a valve.

Trade Terms

Body
Bonnet
Direct pressure
Disc
Packing
Port
Reverse pressure
Seat
Stem
Trim

Required Trainee Materials

1. Pencil and paper
2. Appropriate personal protective equipment

Prerequisites

Before you begin this module, it is recommended that you successfully complete *Core Curriculum*; *Pipefitting Level One*; *Pipefitting Level Two*; *Pipefitting Level Three*; and *Pipefitting Level Four*, Modules 08401-07 through 08407-07.

This course map shows all of the modules in the fourth level of the Pipefitting curriculum. The suggested training order begins at the bottom and proceeds up. Skill levels increase as you advance on the course map. The local Training Program Sponsor may adjust the training order.

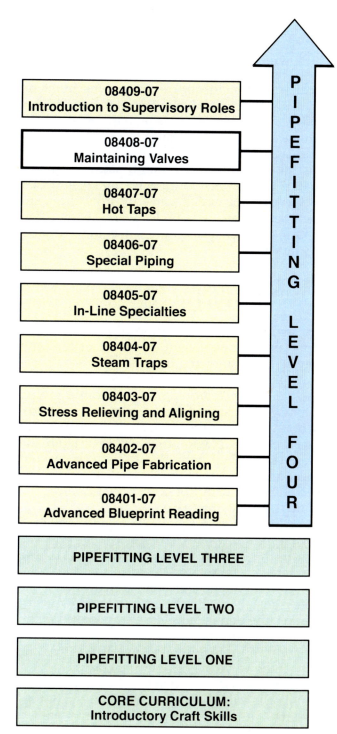

1.0.0 ♦ INTRODUCTION

Valves regulate and direct the flow of material within a piping system. The overall efficiency of any piping system is closely linked to its valves. Because of their importance, valves must be installed correctly and maintained regularly. This module introduces the trainee to the general procedures of valve maintenance.

When maintaining valves on the job site, you must closely follow the manufacturer's maintenance procedures and always make sure that the valve is not under warranty from the manufacturer. If a valve is under warranty and you perform maintenance on the valve, the warranty may be voided.

2.0.0 ♦ REMOVING AND INSTALLING VALVES

Safe removal and installation of valves is very important to the efficient operation of a piping system. Because of regular maintenance procedures, the valve must be easily accessible when possible. When removing valves from a piping system, you must thoroughly follow all plant procedures to lock out the valve and relieve pressure from the valve. If you are working at a job site where you are not permitted to perform lockout procedures, you must ensure that the valve has been locked out by the appropriate personnel before removing the valve.

The location of a valve in a piping system is very important. Valves that are installed with the **stem** in the upright position tend to work best. The stem can be rotated down to the horizontal position but should not point downward. If the valve is installed with the stem in the downward position, the **bonnet** acts like a trap for sediment, which may cut and damage the stem. Also, if water is trapped in the bonnet in cold weather, the **body** of the valve may freeze and crack.

You must also consider the direction of flow through the valve when installing valves. Butterfly, safety, pressure-relief, and some other valves, if they are not bidirectional, either have arrows indicating direction of flow stamped on the side or have **ports** labeled as the inlet or the outlet. When the valve is not marked, you must determine which side of the **disc** you want the pressure against most often. The **trim** of the valve includes the stem disc, seat ring, disc holder or guide, wedge, and bushings; this is the part of the valve that receives the most wear and tear. Most gate valves can have the pressure on either side, as can some butterfly valves. Globe valves should be installed so that the pressure is below the disc unless pressure above the disc is required in the job specifications. *Figure 1* shows pressure above and below the disc.

Tapered plug valves, including multi-port plug valves, are usually installed with the flow pushing in the direction through the plug into the **seat** at any port. This is called **direct pressure**. When the plug is pushed away from the seat, the flow direction is called **reverse pressure**.

Eccentric plug valves used in air, gas, or clean liquid service, are best mounted with the shaft horizontal and in a direct pressure orientation. This means that flow will enter at the end farthest from the seat, and it will push the plug into the seat. If the eccentric plug is on the discharge side of a pump, the plug should be in reverse pressure orientation. If eccentric plugs are used with material such as pulp or other suspended solids, the flow must be able to carry the solids through, without clogging the valve. For that reason, in horizontal pipelines, the plug should open upward. That is, the plug moves away from the seat to the top of the pipe channel. In vertical pipelines the seat end (usually marked either on the body or the flange) is installed at the top, so the plug won't catch or jam on the solids. On DeZurik® eccentric plug valves, the word "Seat" is cast on the seat end of the body. On Valmatic's Cam-Centric® valves, the words "Seat End" are on the flange.

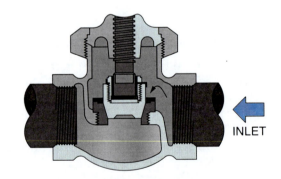

PRESSURE ABOVE DISK

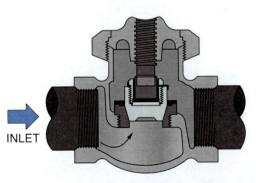

PRESSURE BELOW DISK

Figure 1 ♦ Pressure above and below disc.

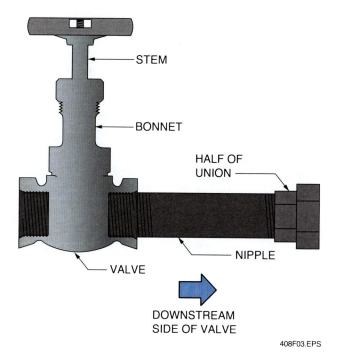

Figure 3 ◆ Completed valve assembly.

when connecting to manufactured equipment. If you must connect a flat-face flange to a raised-face flange, you must confirm that the flat-face flange is not cast iron because the pressure of the bolts can crack or break the cast iron flange. If the flat-face flange is cast iron, change the raised-face flange to a flat-face flange with engineering approval and use a full-face gasket between them.

Flange bolts must be tightened carefully to avoid warping the flange. If the bolts are not tightened correctly, the joint may leak or crack. Pressure must be applied evenly and in the correct amount to make a good seal. A torque wrench can be used to apply equal pressure at all points around the flange. With the aid of a torque wrench, a pipefitter can properly adjust and tighten a flanged joint. Each piping system may use different torque settings, depending on the sizes of the flanges and bolts, the pressure or temperature of the system, and the type of gasket used. Consult your supervisor or the piping drawings for the proper torque setting.

Follow these guidelines when tightening flange bolts, using a torque wrench:

- Before torquing, ensure that the threads of the flange, nut, and bolt are cleaned and lightly oiled or coated with pipe dope or other compounds, if they are recommended by the engineering specifications. Also lubricate the land or face of the nut that is tightened against the flange.
- If the bolts of a flange have already been tightened, loosen and retighten them. Loosen each bolt one full turn, using an open-end or socket wrench.
- Ensure that the torque wrench is correctly adjusted. Most companies calibrate precision tools at regular intervals.
- Always place the torque wrench on the nut, not on the bolt. Use an open-end, box, or socket wrench to hold the bolt while torquing.
- Always use a smooth, even motion when using the torque wrench. A hurried, uneven motion does not give a correct measurement.

CAUTION

Do not begin torquing bolts to the full pressure the first time around the system. This causes unequal pressure on the flanges.

- Use the crossover method of tightening the bolts (*Figure 4*).

Follow these steps to remove and install a flanged valve:

Step 1 Isolate the system according to plant procedures.

Step 2 Perform the standard lockout/tagout procedure.

Step 3 Open the designated relief valves to bleed the system of all pressure.

CAUTION

To prevent damage to the system, ensure that the pipe is supported on both sides of the valve before removing the valve.

Step 4 Break the bolts loose on the far lower side using the correct size wrenches.

Step 5 Break the bolts loose on the far top side using the correct size wrenches.

NOTE

Steps 4 and 5 must be done in this order to ensure that process blast blows away from you.

Step 6 Remove all the bolts from the flanges.

MODULE 08408-07 ◆ MAINTAINING VALVES 8.5

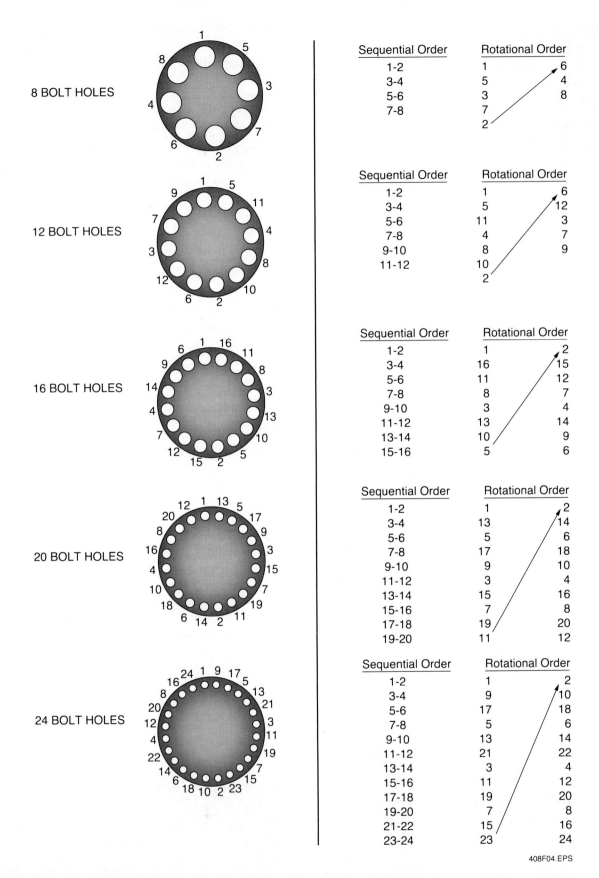

Figure 4 ◆ Crossover method of tightening flange bolts.

8.6 PIPEFITTING ◆ LEVEL FOUR

Step 7 Remove the valve from the system. Flange spreaders may be needed to free the pipe flanges from the valve flanges.

CAUTION
Be sure to use the proper rigging and lifting procedures to eliminate the possibility of damaging the valve.

Step 8 Scrape the gasket surfaces clean using a sharp putty knife.

WARNING!
If the gasket has deteriorated, it may be made of asbestos. You must take special precautions when working with asbestos. Always check with your supervisor if you are unsure of the gasket material.

Step 9 Inspect the flanges for cuts, grooves, or damage.

CAUTION
Be careful when scraping the gasket surfaces to avoid damaging the flange face of the valve.

Step 10 Move the valve to the valve repair area.

Step 11 Select the proper replacement valve to be installed into the system according to specifications.

Step 12 If specified, apply antiseize compound or other materials to the bolts and gaskets.

Step 13 Maneuver the valve into the proper position, using proper rigging procedures.

Step 14 Align the bolt holes between the flange on the valve and the pipe in the system using a drift pin.

WARNING!
Never use your fingers to align flange holes.

NOTE
If the flanges are flat-faced flanges, insert the full-face gaskets between the flanges before inserting the flange bolts. Hold the full-face gaskets in place, using the drift pins.

Step 15 Place the bolts in the bottom bolt holes.

Step 16 Put the nuts on the bolts loosely.

NOTE
If using raised-faced flanges, slide the gaskets into position so that the bolts on the bottom cradle the gaskets and keep them in place.

Step 17 Install all the bolts and nuts in the holes in the flange.

Step 18 Ensure that the valve is square, using a level.

NOTE
Make sure that the flanges mate evenly. Check the manufacturer's specifications for the torque on the bolts.

Step 19 Tighten the bolts to 25 percent of the required torque using the crossover method.

Step 20 Tighten the bolts to 50 percent of the required torque using the crossover method.

Step 21 Continue tightening the bolts using the crossover method and increasing the torque by 25 percent until the required torque has been reached. Then repeat until the movement stops. If movement occurs, loosen all bolts and start over.

2.3.0 Troubleshooting Valves

Operate the valve a few times to see if the mechanism is working. If the actuator is very difficult to operate, there are several possible reasons:

The packing may be too tightly compressed. Loosen the packing nut, if the contents are not too dangerous, until it leaks slightly, then tighten until the leak stops.

Dirt may be obstructing the operation of the valve. This is frequently a problem in ball valves especially, because of the way the valve opens and closes. The usual solution is to flush the valve out.

In the case of gate valves, and to a lesser extent, globe valves, if the valve won't open, there may be pressure above the seat. If there is an equalizer on the valve, that might solve the problem.

3.0.0 ◆ VALVE STEMS AND O-RINGS

The valve stem provides a link between the handwheel and the gate. Many times, the pipefitter must attach the handwheel to the stem after the valve is installed. Some handwheels bolt onto the stem; others slip over a keyway cut into the stem and are then secured to the stem by a nut or machine screw. Handwheels vary according to the type of stem they fit. Three general types of stems include the rising stem, the nonrising stem, and the outside screw-and-yoke (OS&Y) stem. *Figure 5* shows types of stems.

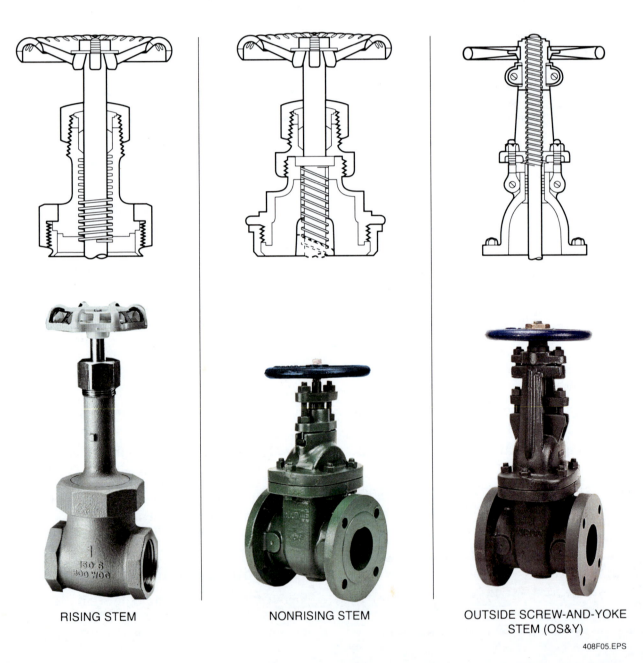

RISING STEM

NONRISING STEM

OUTSIDE SCREW-AND-YOKE STEM (OS&Y)

Figure 5 ◆ Stems.

When a rising stem valve is opened, both the handwheel and the stem rise. The height of the stem gives an approximation of how far the valve is open. These stems can be installed only where there is sufficient clearance for the handwheel to rise. Rising stems come in contact with the fluid in the valve and are used only with fluids that do not harm the threads, such as hydrocarbons, water, and steam.

Nonrising stems are suitable in spaces where space is limited. As the name implies, neither the stem nor the handwheel rises when the valve is opened. A spindle inside the valve body turns when the handwheel is turned and raises the stem; therefore, stem wear is held to a minimum. This type of stem also comes in contact with the fluid in the valve.

The OS&Y stem is suitable for corrosive fluids because the stem does not come in contact with the fluid in the line. As the handwheel is turned, the stem moves up through the handwheel. The height of the stem indicates how far the valve is open.

Some valve stems have O-rings that provide a leakproof seal for the lubrication around the stem. These O-rings fit inside specially machined grooves. *Figure 6* shows O-rings on a valve stem.

3.1.0 Types of O-Rings

O-rings (*Figure 7*) are rugged and extremely dependable. They are used to seal against conditions ranging from strong vacuum to high pressure. O-rings are made from a variety of materials for different applications. Like gaskets, O-rings are often used to seal a mechanical connection between two parts of an instrument. O-rings that are used in instruments are often made of rubber or of a synthetic material. Occasionally, high-temperature or pressure applications may require the use of a metal O-ring. Among the characteristics useful in O-rings is low compression set, that is, the ability to return to their original shape and size when they are released from pressure.

Some of the more common O-rings are made from Buna-N (Nitrile), ethylene propylene, Viton®, Teflon®, silicone, Teflon®-encapsulated silicone, and polyurethane. O-rings are used in both static and dynamic seals. A static seal is not subjected to flow, but may have system pressure. A dynamic seal has both flow and pressure.

The sizes of O-rings are set by the *Aerospace Standard AS568A*, published by the Society of Automotive Engineers (SAE). Sizes are designated by dash numbers, such as AS568-006 and AS568-216.

3.1.1 Buna-N O-Rings

Buna-N O-rings are widely used. They are made of an elastomeric-sealing material. They are used with a variety of petroleum and silicone fluids, hydraulic and nonaromatic fuels, and solvents. Buna-N O-rings are not compatible with phosphate esters, ketones, brake fluids, strong acids, or ozone. They have an approximate temperature range of –65°F to 275°F. Buna-N O-rings do not weather well, especially in direct sunlight.

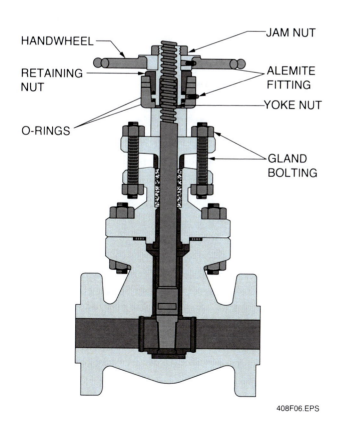

Figure 6 ◆ O-rings on a valve stem.

Figure 7 ◆ O-rings.

3.1.2 Ethylene Propylene O-Rings

Ethylene propylene O-rings resist automotive brake fluids, hot water, steam to approximately 400°F, silicone fluids, dilute acids, and phosphate esters. They are not compatible with petroleum fluids and diester lubricants. Ethylene O-rings have a good compression set plus high abrasion resistance, and they are weather resistant. They have an approximate temperature range of –70°F to 250°F.

3.1.3 Viton® O-Rings

Viton® O-rings offer excellent resistance to petroleum products, diester lubricants, silicone fluids, phosphate esters, solvents, and acids, except fuming nitric acid. They have a low compression set and low gas permeability. Viton® O-rings are often used for hard vacuum service. They should never be used with acetates, methyl alcohol, ketones, brake fluids, hot water, or steam. The approximate temperature range of Viton® O-rings is –31°F to 400°F.

3.1.4 Teflon® O-Rings

Teflon® O-rings lubricate well and have excellent chemical and temperature resistance. They make fine static seals, but need mechanical loading when used as dynamic seals. They have an approximate temperature range of –300°F to 500°F.

NOTE

All O-rings require lubrication. Keep in mind, however, that the lubricant and the O-ring material must be compatible. If they are not properly matched, the seal could be damaged. Because new lubricants are constantly being introduced, it is increasingly important to check the manufacturer's requirements to ensure a match between the O-ring and the lubricant.

3.1.5 Silicone O-Rings

Silicone O-rings are used where long-term exposure to dry heat is expected. Due to poor abrasion resistance, silicone O-rings perform best in static sealing applications. Silicone O-rings resist brake fluids and high aniline point oil. They are not recommended for use with ketones and most petroleum oils. They have an approximate temperature range of –80° to 400°F.

3.1.6 Teflon®-Encapsulated Silicone O-Rings

Teflon®-encapsulated silicone O-rings are used in most of the same applications as silicone O-rings. The Teflon® coating makes them resistant to most solvents and chemicals. These O-rings have an extremely low coefficient of friction and low compression set. They are primarily used as seals in static applications. They have an approximate temperature range of –75°F to 400°F.

3.1.7 Polyurethane O-Rings

Polyurethane O-rings have high tensile strength, are abrasion resistant, and have excellent tear strength. Polyurethane is the toughest of the elastomers. Polyurethane O-rings can be used with petroleum fluids, ozone, and solvents, except ketones. Polyurethane O-rings are noncompatible with hot water, brake fluids, acids, and high temperatures. They have poor compression set. Polyurethane O-rings have an approximate temperature range of –40°F to 200°F.

3.2.0 Replacing Valve Stem O-Rings

Follow these steps to replace valve stem O-rings:

Step 1 — Remove the jam nut from the valve stem.

Step 2 — Remove the handwheel from the valve stem.

Step 3 — Remove the retaining nut, using a box wrench.

Step 4 — Screw the yoke nut out of the top of the bonnet.

Step 5 — Remove and inspect the O-rings, looking for any damage caused by the valve stem.

Step 6 — Smooth the surface on the valve stem, using a fine-grit emery cloth.

Step 7 — Select the correct O-ring replacements. The manufacturer's specifications specify the proper material and size O-rings to use. The O-rings must be the correct size and made of the proper material to provide an adequate seal in the valve.

Step 8 — Lubricate the O-rings lightly, according to specifications, to prepare them for installation.

Step 9 — Slide the O-rings into the appropriate grooves.

Step 10 — Screw the yoke nut onto the valve stem very carefully.

CAUTION

To avoid breaking the seal, do not pinch or break the O-rings, and do not dent or scratch the valve stem with the bonnet.

Step 11 Screw the retaining nut onto the yoke nut.

Step 12 Return the handwheel to the valve stem.

Step 13 Return the jam nut to the valve stem.

Step 14 Tighten the jam nut.

Step 15 Open and close the valve several times to make sure it is working properly. If the valve is not working properly, repeat Steps 1 through 15; otherwise, proceed to Step 16.

Step 16 Return the valve to service.

4.0.0 ◆ REPLACING BONNET GASKETS

The types of bonnets that contain gaskets or seals are the bolted, union, or threaded bonnets (*Figure 8*). Bolted bonnets are bolted to the body by several bolts around the circumference of the bonnet. Valves with bolted bonnets are commonly used for line sizes larger than 1¼ inches, and are designed for use in process plants at all pressures and temperatures. When a leak occurs between the bonnet and the body of the valve, you must replace the bonnet gasket.

Threaded bonnets have a male threaded piece that screws inside the body of the valve, with the stem passing through the center of the male threaded top. A union bonnet has a female threaded ring or nut screws onto the outside of the body of the valve. A bolted bonnet may involve

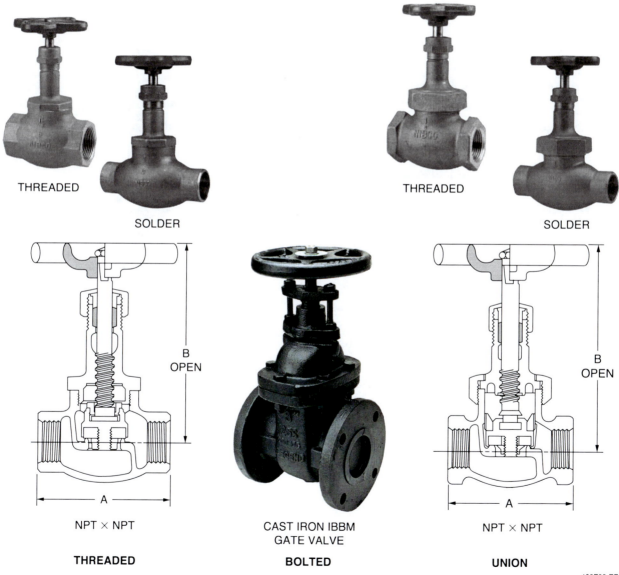

Figure 8 ◆ Bolted, union, and threaded bonnets.

bolts screwing into threaded holes in the body or flanges, with holes all the way through and nuts providing the threads to tighten the flanges together.

Follow these steps to replace bolted bonnet gaskets:

Step 1 Isolate the system according to plant procedures.

Step 2 Perform the standard lockout/tagout procedure. Follow your company's line break procedure.

Step 3 Remove the valve from the system so that testing can be performed after replacing the gasket.

Step 4 Secure the valve in a vise, with the valve ends between the jaws of the vise.

CAUTION

Use soft covers in the vise jaws to protect the valve from damage.

Step 5 Turn the handwheel to open the gate valve (*Figure 9*). This raises the disc from the valve seat into the bonnet area.

Step 6 Remove the bonnet bolts from the bonnet.

Step 7 Remove the bonnet from the valve body.

Step 8 Remove the gasket and all gasket material remaining on the bonnet using a putty knife, gasket scraper, or single-edge razor blade. When removing the old gasket, try to keep it in one or two pieces so that it can be used as a pattern for a new gasket.

Step 9 Select the proper gasket material to make a new gasket. Always use the manufacturer's specifications to determine the proper material to use. Certain types of gaskets cannot be made on site and must be ordered from the manufacturer.

Step 10 Place the old gasket on top of the gasket material, and trace the outline using a pencil. If the old gasket is completely destroyed, use the bonnet as a pattern for the new gasket.

Step 11 Cut out the gasket using a gasket cutter.

Step 12 You may or may not lubricate the gasket, depending on the specifications.

Step 13 Place the gasket on the valve, making sure it is positioned properly.

Step 14 Place the bonnet on the valve body.

Step 15 Clean and lubricate the bonnet bolt threads.

Step 16 Place the bonnet bolts in the proper position.

Step 17 Hand-tighten the bonnet bolts.

Step 18 Tighten the bonnet bolts using the crossover method. Follow the manufacturer's specifications, and use a torque wrench to tighten the bolts.

Step 19 Replace the handwheel and tighten it according to specifications.

Step 20 Test the valve according to the specified test procedure.

Step 21 Return the valve to service.

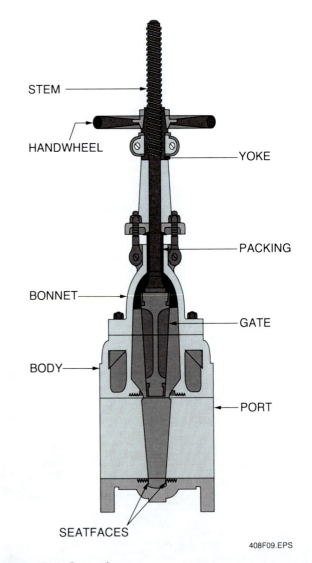

Figure 9 ◆ Gate valve.

5.0.0 ◆ PACKING VALVES

There are four areas within a valve that must provide a complete seal if the valve is to function properly. These areas are as follows:

- The closing mechanism inside the valve
- The end connections
- Where the valve bonnet meets the body
- The valve stem

Achieving a good seal at the stem is somewhat difficult because of the movement of the stem. Therefore, various types of packing are used to seal the stem. Packing prevents fluid from leaking up through the valve stem. It also retains the pressure of the fluid within the valve. The packing material fills the stuffing box, which is the space between the valve stem and the bonnet. (See *Figure 10.*) A gland presses the packing against the stem within the stuffing box. The stuffing box usually requires occasional tightening, especially if the valve has not been used for a while.

5.1.0 Packing Shapes and Materials

The most common types of packing are solid, braided, and granulated fibers. Packing also takes many different shapes. It may be square, wedge-shaped, or ring-shaped. Commonly used types of packing include Teflon® yarn and filament, lubricated graphite yarn, lubricated carbon yarn (graphite impregnated), and tetrafluoroethylene (TFE/synthetic fiber). Types of packing are shown in *Figure 11*. Though most packing materials will normally be specified, the following factors must be considered when choosing the type of packing material to use:

- Material flowing through the line
- Operating pressures
- Operating temperatures
- Minimum temperature of the piping system
- Composition of the valve stem

5.1.1 Teflon® Yarn Packing

Teflon® yarn packing is a cross-braided yarn impregnated with Teflon®. Teflon®-impregnated packing resists concentrated acids, such as sulfuric acid and nitric acid, sodium hydroxide, gases, alkalis, and most solvents. It is good for use in applications of up to approximately 550°F.

5.1.2 Teflon® Filament Packing

Teflon® filament (cord) packing is a braided packing made from TFE filament. It is impregnated with Teflon® and an inert softener lubricant, or sometimes with graphite. It is often used on rotating pumps, mixers, agitators, kettles, and other equipment.

5.1.3 Lubricated Graphite Yarn Packing

Lubricated graphite yarn packing is an intertwined braid of pure graphite yarn impregnated with inorganic graphite particles. The graphite particles dissipate heat. Lubricated graphite yarn packing also contains a special lubricant that provides a film to prevent wicking and reduce friction. It is good for use in high-temperature applications.

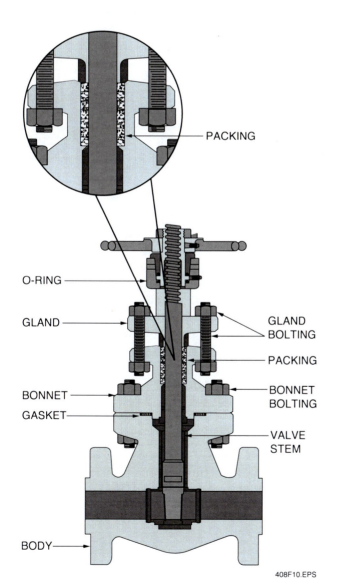

Figure 10 ◆ Packing.

INCONEL® AND CARBON WIRE YARN

TEFLON® YARN

GRAPHITE YARN

GRAPHITE RIBBON

PACKING IN PLACE

SYNTHETIC FIBER

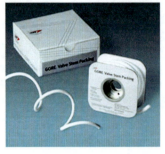
EXPANDED PTFE

Figure 11 ♦ Types of packing.

5.1.4 Lubricated Carbon Yarn Packing

Lubricated carbon yarn (graphite impregnated) packing is made from an intertwined braid of carbon fibers impregnated with graphite particles and lubricants to fill voids and block leakage. Lubricated carbon yarn packing is used in systems containing water, steam, and solutions of acids and alkalis. It is considered suitable for steam applications with temperatures up to approximately 1,200°F, and up to 600°F where oxygen is present. It is reactive to oxygen atmospheres.

5.1.5 TFE/Synthetic Fiber Packing

TFE/synthetic fiber packing is made from braided yarn fibers saturated and sealed with TFE particles before being woven into a multi-lock braided packing. TFE/synthetic fiber packing protects against a variety of chemical actions. It is used in applications where caustics, mild acids, gases, and many chemicals and solvents are present. It is often used in general service for rotating and reciprocating pumps, agitators, and valves.

Some variations also use TFE yarns, sometimes combined with Inconel, and a filling of graphite or other material, held in place with the yarn exterior.

5.2.0 Repacking Valves

If a valve is leaking through the stem, tightening the valve usually stops the leak. If tightening does not stop the leak, the packing may need to be checked.

Always store packing materials in a clean, dry place, leaving them in the original wrappings. Never allow dirt or abrasive materials to come in contact with the packing material.

The precise procedure for replacing packing varies according to the type of valve and the type of packing. The following is a general procedure for repacking a valve:

Step 1 Isolate the system according to plant procedures. The first requirement is that the valve must be fully backseatable and must be backseated.

WARNING!

Never repack a valve with the valve under pressure unless specifically told to do so by your supervisor. The circumstances in which you would do this will not be a judgment call of yours. The first requirement is that the valve be completely backseated; that is, the valve is made so that it can be opened all the way to seal the stem and packing completely from the pressure of the fluid. This is still a dangerous operation, especially with high temperature or reactive materials.

Step 2 Perform the standard lockout/tagout procedure.

Step 3 Isolate the valve and remove it from the system.

Step 4 Secure the valve in a vise, with the valve ends between the jaws of the vise.

CAUTION

Use soft covers in the vise jaws to protect the valve from damage.

Step 5 Turn the handwheel to open the valve completely.

Step 6 Remove the handwheel nut from the valve stem.

Step 7 Remove the handwheel from the valve stem.

Step 8 Remove the gland bolts.

Step 9 Slide the gland flange off the valve body and over the stem.

Step 10 Slide the packing gland off the valve stem, exposing the valve packing area.

Step 11 Remove the valve packing material, using a valve packing removal tool (*Figure 12*).

Step 12 Clean the stem, packing gland, and stuffing box using a fine-grit emery cloth, making sure all traces of the old packing material are removed.

Step 13 Select new packing material recommended by the manufacturer or plant specifications.

Step 14 Wrap a length of packing around the valve stem.

Step 15 Mark two 45-degree lines where the packing material overlaps. *Figure 13* shows marking the valve packing material.

Step 16 Remove the packing from the stem.

Step 17 Cut the packing along the 45-degree marks on the packing material using a knife.

CAUTION

Do not cut the packing on the stem because this will damage the stem.

Step 18 Place the cut piece of packing around the stem to check the fit. The packing material should fit snugly around the stem.

Step 19 Remove the packing from the stem. This piece is the pattern for cutting more pieces of packing.

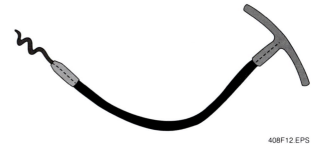

Figure 12 ♦ Valve packing removal tool.

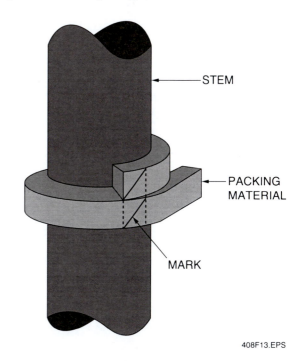

Figure 13 ♦ Marking valve packing material.

Step 20 Cut several pieces of the packing material, using the pattern.

Step 21 Push the first piece of packing into the stuffing box using the packing gland to push it down.

Step 22 Place another piece of packing in the stuffing box using the packing gland to push it down. Rotate the alignment of the packing joints 180 degrees with each layer so they do not line up (*Figure 14*).

Step 23 Continue stuffing the packing one piece at a time until the stuffing box is full.

Step 24 Slide the packing gland over the stem.

Step 25 Slide the gland flange over the stem and onto the valve body.

Step 26 Replace and handtighten the gland bolts.

Step 27 Replace the handwheel on the stem.

Step 28 Replace and tighten the handwheel nut.

Step 29 Clean and lubricate the gland bolt threads.

Step 30 Insert the gland bolts into the glands.

Step 31 Tighten all bolts using the crossover method. Use a torque wrench and follow the valve manufacturer's specifications for the proper torque.

Step 32 Return the valve to service.

Step 33 Place a few drops of oil on the stem, depending on the specifications.

Step 34 Turn the handwheel a few times to ensure that it is turning properly. If it is not turning properly, repeat Steps 6 through 32; otherwise, proceed to Step 34.

Step 35 Test the valve if required.

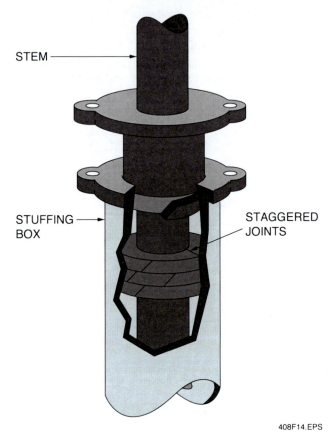

Figure 14 ◆ Staggered packing joints.

Review Questions

1. When the plug of a valve is pushed away from the seat, this results in _____.
   a. direct pressure
   b. a reverse pressure
   c. a venturi port
   d. a full port

2. Which part of the valve receives the most wear and tear?
   a. Trim
   b. Port
   c. Stem
   d. O-ring

3. When installing a threaded valve, you will most often lubricate the _____.
   a. threads of the pipe
   b. threads inside the valve
   c. inside of the pipe
   d. innermost thread of the pipe in the valve

4. The first step of installing a threaded valve is to _____.
   a. isolate that part of the system
   b. cut the pipe on the downstream side
   c. open the drain valves
   d. make sure the pressure is on at that valve

5. When installing flanged valves, avoid unequal pressure on the flanges by using the crossover method of tightening flange bolts and _____.
   a. torquing bolts to the full pressure the first time around
   b. increasing torque by 25 percent each time around until the required torque has been reached
   c. increasing torque by 50 percent each time around until the required torque has been reached
   d. placing the torque wrench on the bolt, not the nut

6. What type of stem should be used with corrosive fluids?
   a. The OS&Y
   b. Silicone O-ring
   c. Rising stem
   d. Non-rising stem

7. What type of O-ring does *not* weather well?
   a. Polyurethane
   b. Silicone
   c. Teflon®
   d. Buna-N

8. Valves with bolted bonnets are commonly used for line sizes larger than _____ inch(es).
   a. ⅛
   b. ¼
   c. ½
   d. 1¼

9. All of the following must be considered when choosing packing for a valve *except* _____.
   a. material flowing through the line
   b. operating pressure
   c. whether the valve is threaded
   d. the composition of the valve stem

10. The ends of the packing should be _____.
    a. cut square
    b. staggered
    c. curved
    d. straight up and down

Summary

Proper storage, installation, and maintenance of valves and all piping components ensure that a system is operating as safely and efficiently as possible. Pipefitters are actively involved in the maintenance of valves to keep the piping system in proper working order. You must be aware of the various types of sealants and packing materials in order to safely select the right one for given pressure, temperature, and environmental conditions. Before maintaining any valve, be sure that you follow the proper procedures to isolate the valve and relieve pressure from the valve. Placing safety first in all maintenance activities ensures your safety and the safety of your co-workers.

Notes

Trade Terms Introduced in This Module

Body: The main part of the valve. It contains the disc, seat, and valve ports. The body of the valve is directly connected to the piping by threaded, welded, mechanically joined, or flanged ends.

Bonnet: The part of the valve that contains the trim. The bonnet is located above the body.

Direct pressure: Flow pushing the sealing element of the valve into the seat and improving closure.

Disc: The moving part of a valve that directly affects the flow through the valve.

Packing: The material between the valve stem and bonnet that provides a leakproof seal and prevents material from leaking up around the valve stem.

Port: The internal opening for flow through a valve.

Reverse pressure: Flow pushing the closure element out of the seat.

Seat: The nonmoving part of the valve on which the disc rests to form a seal and close off the valve.

Stem: The part of the valve that connects the disc to the valve operator. A stem can have linear, rotary, or helical movement.

Trim: The internal parts of a valve that receive the most wear and can be replaced. The trim includes the stem disc, seat ring, disc holder or guide, wedge, and bushings.

Resources & Acknowledgments

Additional Resources

This module is intended to present thorough resources for task training. The following reference works are suggested for further study. These are optional materials for continued education rather than for task training.

Choosing the Right Valve. New York, NY: Crane Company.

Piping Pointers; Application and Maintenance of Valves and Piping Equipment. New York, NY: Crane Company.

http://www.dezurikwater.com/basic_valves_instruction_index.htm

http://www.valmatic.com/manuals.jsp

http://www.velan.com/products/index.htm

http://www.acipco.com/

Figure Credits

RED-WHITE Valve Corp., 408F05 (bottom left)

NIBCO INC., 408F05 (bottom left, bottom right), 408F07, 408F08 (threaded valve, union valve)

Legend Valve and Fitting, Inc., 408F08 (center)

Sealing Equipment Products Co., Inc. (SEPCO®), 408F11A–408F11E

A.W. Chesterton, 408F11F

GORE™ Valve Stem Packing © W.L. Gore & Associates, 408F11G

NCCER CURRICULA — USER UPDATE

NCCER makes every effort to keep its textbooks up-to-date and free of technical errors. We appreciate your help in this process. If you find an error, a typographical mistake, or an inaccuracy in NCCER's curricula, please fill out this form (or a photocopy), or complete the online form at **www.nccer.org/olf**. Be sure to include the exact module ID number, page number, a detailed description, and your recommended correction. Your input will be brought to the attention of the Authoring Team. Thank you for your assistance.

Instructors – If you have an idea for improving this textbook, or have found that additional materials were necessary to teach this module effectively, please let us know so that we may present your suggestions to the Authoring Team.

NCCER Product Development and Revision
13614 Progress Blvd., Alachua, FL 32615

Email: curriculum@nccer.org
Online: www.nccer.org/olf

❑ Trainee Guide ❑ AIG ❑ Exam ❑ PowerPoints Other _____

Craft / Level: _____ Copyright Date: _____

Module ID Number / Title: _____

Section Number(s): _____

Description: _____

_____

_____

_____

_____

Recommended Correction: _____

_____

_____

_____

_____

Your Name: _____

Address: _____

_____

Email: _____ Phone: _____

Pipefitting Level Four

08409-07

Introduction to Supervisory Roles

08409-07
Introduction to Supervisory Roles

Topics to be presented in this module include:

1.0.0   Introduction to Supervision 9.2
2.0.0   The Construction Industry Today 9.3
3.0.0   Gender and Minority Issues 9.4
4.0.0   Construction Projects 9.7
5.0.0   The Construction Organization 9.7
6.0.0   Becoming a Leader 9.9
7.0.0   Problem Solving and Decision Making 9.13
8.0.0   Safety Responsibilities 9.17

Overview

Many pipefitters have no desire to move into supervision. However, those who do choose to move beyond the pure craftwork level have many things to learn. This module describes the necessary skills for the pipefitter who wants to move into supervision. The issues covered include communication, issues of cultural and gender differences, and legal and ethical questions. This module, although hardly a complete training for the supervisory levels, is a presentation of the many lessons that the trainee will need to pursue in order to become a supervisor.

Objectives

When you have completed this module, you will be able to do the following:

1. Explain the importance of training for construction industry personnel.
2. Identify the gender and minority issues associated with a changing workforce.
3. Describe what employers can do to prevent workplace discrimination.
4. Describe the four major categories of construction projects.
5. Describe the difference between formal and informal organizations, as well as the difference between authority and responsibility.
6. Explain the purpose and content of a job description and a policy/procedure document.
7. List the characteristics and behavior of effective leaders, as well as the different leadership styles.
8. Explain the difference between problem solving and decision making.
9. Describe strategies for reducing absenteeism and turnover.
10. Explain the duties of a crew leader in enforcing safety on the job.

Required Trainee Materials

1. Pencil and paper
2. Appropriate personal protective equipment

Prerequisites

Before you begin this module, it is recommended that you successfully complete *Core Curriculum*; *Pipefitting Level One*; *Pipefitting Level Two*; *Pipefitting Level Three*; and *Pipefitting Level Four*, Modules 08401-07 through 08408-07.

This course map shows all of the modules in the fourth level of the Pipefitting curriculum. The suggested training order begins at the bottom and proceeds up. Skill levels increase as you advance on the course map. The local Training Program Sponsor may adjust the training order.

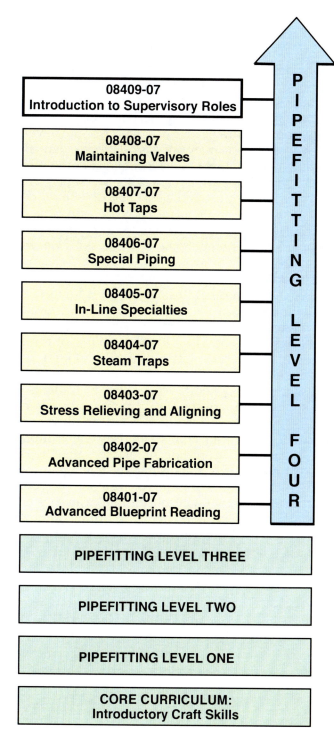

1.0.0 ♦ INTRODUCTION TO SUPERVISION

Some workers have no desire to progress beyond being an effective and successful craftworker. That is not a defect or a fault; if you feel content with your position, you may be happier as a craftworker. Supervision is not for everyone. For those who wish to become a crew leader or supervisor, however, there are a great many new skills to be learned. The knowledge that you have attained of the craft itself is certainly part of those skills. The supervisor or crew leader must know the skills of their workers as well, in order to properly supervise their work. However, supervision requires a great many people skills as well. Some of those skills can be acquired simply by working with others in teams. Other skills may require further training.

For the purpose of this module, it is important to define some of the positions to which we will be referring. Craftworkers and supervisors of all ages have depended on the availability of three elements: materials, labor, and technology. The term craftworker refers to a person who performs the work of his or her trade(s). The crew leader is a person who supervises one or more craftworkers on a crew. A superintendent is essentially a second-line supervisor who supervises one or more crew leaders or front-line supervisors. Finally, a project manager is responsible for managing the construction of one or more construction projects. In this module, we will concentrate primarily on the role of the crew leader. The organizational chart in *Figure 1* depicts how each of the four positions is related in the chain of command.

Craftworkers and crew leaders differ in that the crew leader manages the activities that the skilled craftworkers on the crews actually perform. In order to manage a crew of craftworkers, you must have first-hand knowledge and experience in the activities being performed. In addition, you must be able to act directly in organizing and directing the activities of the various crew members.

Some basic skills of the supervisor include, but are not limited to, the following:

- Determining the training needs of workers and training them
- Maintaining good relationships with and among workers
- Dealing with workers who are not from the same culture, or who are of different genders
- Solving personal problems between workers
- Budgeting time and materials effectively
- Motivating workers to perform efficiently

This module discusses the importance of developing effective leadership skills as a new supervisor and provides some tips for doing so. Effective ways to communicate with all levels of employees, build teams, motivate crew members, make decisions, and resolve problems will be covered in depth. You will learn about the issues you will need to be aware of as a supervisor and how you can maintain good morale and productivity in your team of workers. Once you have completed this module, you will know how to manage a diverse workforce and maintain high standards of training, communication, and ethics.

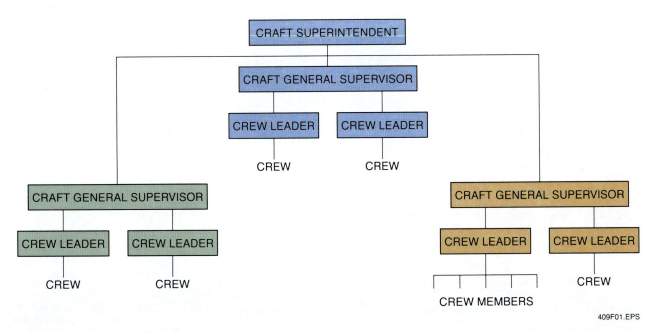

Figure 1 ♦ Project organization chart.

2.0.0 ♦ THE CONSTRUCTION INDUSTRY TODAY

There have been many changes in the construction industry over the years. The continued need for training and education, advances in technology, and the addition of women and minorities to the workforce have affected the way that construction companies operate. The three best sales tools used by the construction industry to attract qualified job candidates are training and education; new technologies; and career growth.

2.1.0 Training

The need for continuing craft training is evident if the industry is to meet the forecast demand for workers. The Construction Labor Research Council estimated that the construction industry would need more than 240,000 new workers annually, beginning in the year 2001. Replacement needs will probably exceed growth demands by 250 percent.

Because older workers in construction are retiring earlier and younger workers are more likely to get jobs in other industries, the construction industry has been forced to increase its share of the new job entrants to meet labor demands. The Department of Labor (DOL) concludes that the best way for the construction industry to reduce shortages of skilled workers is to create more education and training opportunities. The DOL suggests that companies and community groups form partnerships and create apprenticeship programs. Such programs could provide women, young people, and minorities an opportunity to develop the skills needed for construction work by giving them hands-on experience.

Present construction craft apprenticeship and task-training programs are graduating an average of 50,000 people per year. If the training of construction workers is not significantly increased, a severe shortage of skilled help could result. In fact, signs of this shortage have already appeared, as wages for construction work have risen to be an average of $4 more per hour than in other private industries, according to the Bureau of Labor Statistics (BLS). Despite the fact that the BLS shows that the number of workers employed in construction reached an all-time high in 2006, the labor supply still does not satisfy demand. Therefore, the two major concerns of the construction industry remain:

- Developing adequate training programs to provide skilled craftworkers to meet future needs
- Providing supervisory training to ensure there are qualified leaders in the industry to supervise the craftworkers

2.1.1 Craft Training

Most of the craft training done in the construction industry is rather informal, taking place on the job site, outside of a traditional training classroom. According to the American Society for Training and Development, craft training is generally handled through on-the-job instruction by a co-worker or conducted by a supervisor.

The Society of Human Resources Management (SHRM) offers the following tips to supervisors in charge of training their employees:

- *Establish career goals for each person in your area* – This information will give you an idea of how much time you need to devote to developing each person. Employees with the highest ambitions and those with the lowest skill levels will require the most training.
- *Determine what kind of training to give* – Training can be on the job under the supervision of a co-worker, it can be one-on-one with the supervisor, it can involve cross training to teach a new trade or skill, or it can involve delegating new or additional responsibilities.
- *Determine the trainee's preferred method of learning* – Some people learn best by watching, others from verbal instructions, and others from doing. When training more than one person at a time, use all three methods.

Supervisors should do the following when training employees:

- *Explain the task, why it needs to be done, and how it should be done* – Confirm that the trainees understand these three areas by asking questions. Allow them to ask questions as well.
- *Demonstrate the task* – Break the task down into manageable chunks. Do one part at a time.
- *Ask trainees to do the task while you observe them* – Try not to interrupt them while they are doing the task unless they are doing something that is unsafe and potentially harmful.
- *Give the trainees feedback* – Be specific about what they did, and mention any areas where they need to improve.

2.1.2 Supervisory Training

Given the increased demand for skilled craftworkers and qualified supervisory personnel, it seems logical that training would be on the rise.

While some contractors have developed their own in-house training program and others participate in training offered by associations and other organizations, many others do not offer training at all.

If you are a craftworker who wants to move up into supervisory and managerial positions, it will be necessary for you to continue your education and training, especially in the technology that you will need in order to complete your supervisory responsibilities. Technology training includes becoming familiar with computerized record-keeping and estimating, email and internet, and other electronic communications devices. That training should be made fully available to you. Those who are willing to acquire and develop new skills will have the best chance of finding stable employment. It is therefore critical that construction companies remove the barriers against, and devote the needed resources to, quality training.

3.0.0 ◆ GENDER AND MINORITY ISSUES

As a supervisor, you will have to direct all employees who are provided to you. With the trend of increasing diversity in our society, supervisors need to learn how to effectively communicate with people from a variety of cultures, backgrounds, and experiences.

In his book, *Communication: The Essence of Labor Relations*, Richard Hutchinson reported that the number of women in the construction industry increased nearly 16 percent between 1995 and 1999. By the end of 1999, there were approximately 886,000 female construction workers versus 8,101,000 male construction workers in the industry. By 2004, the BLS estimated there were 1,041,000 women in construction versus 9,727,000 men. The United States Census Bureau reported that there are approximately 24 million foreign-born residents who are becoming a greater proportion of the workforce in America.

As women and minorities make up a larger percentage of the construction workforce, some issues relating to gender, race, and ethnicity must be addressed on the job site. These include different communication styles of men and women, language barriers associated with cultural differences, the possibility of sexual harassment, and the potential for gender or racial discrimination.

3.1.0 Communication Styles of Men and Women

As more and more women move into construction, it becomes increasingly important that communication barriers between men and women are broken down and differences in behaviors are understood so that they can work together more effectively.

The Jamestown, New York Area Labor Management Committee (JALMC) offers the following explanations and tips:

- *Women tend to ask more questions than men do* – They are interested, and they want to learn. Men can support their learning by answering their questions and adding a few questions of their own. That way, both genders are involved in the problem-solving and decision-making process.

- *Men tend to offer solutions before empathy; women tend to do the opposite* – Both men and women should say what they want up-front, whether it is a solution or a sympathetic ear. That way, both genders will feel understood and supported.

- *Women are more likely to ask for help than men* – Women should let men know that they can and are willing to help. In addition, men should avoid doing a task for a woman when she asks for help; instead, they should show her how to do it so that she can do it alone the next time. That way, both men and women get the help that they need, and things are accomplished more quickly.

- *Men tend to communicate more competitively, and women tend to communicate more cooperatively* – Women should finish their statements, even if interrupted, and they should remember that interruptions are not necessarily meant to be rude, but are a man's way of asking questions and showing interest. Men should hear the woman out before interrupting. The result is a team or work unit that is both cooperative and competitive, a great combination.

- *To establish trust with one another, women tend to self-disclose while men focus on reliability* – Women should try to keep all commitments with men, even small ones. In the event that they cannot be kept, women should attempt to renegotiate so that a positive track record of reliability is maintained. Men should try to understand that women who share their feelings are doing so because they feel that they can trust the other person, not because they want to appear unstable or weak. Recognizing these two different styles leads to less suspicion and hesitancy about working together, and contributes to a better flow of information.

3.2.0 Language Barriers

Language barriers are a real workplace challenge. The U.S. Census Bureau reported that the number of people who speak languages other than English increased from 23.1 million to 31.8 million between 1980 and 1990. By the year 2000, that number was over 37 million. Spanish is now the most common non-English language spoken in the United States. As of 2006, the BLS estimated that Hispanics made up 30 percent of the overall construction workforce and 8 percent held construction management positions. As the make-up of the immigrant population continues to change, the number of non-English speakers will continue to rise, and the languages being spoken will also vary.

According to an article in *USA Today*, the U.S. population will grow by about 130 million people by the year 2050. About one-third of those people will be from other countries. The article offers the following solutions based on what companies have done to overcome this challenge:

- Hire supervisors who speak the language of the non-English speaking workers. You, as a crew leader, would do well to learn to communicate using the language of your employees.
- Offer English classes either at the work site or through school districts and community colleges.
- Offer incentives for workers to learn English.
- Ask employees who are parts of the problem to become parts of the solution.
- Do not assume that the work group's common language must be English.
- Allow work groups to pick the language that works best for them, except on occasions where a common language may be necessary.
- Develop a policy statement about languages in the workplace.

As our workforce becomes more diverse, communicating with people for whom English is a second language will be even more critical. The Knowledge Center's Manager's Tool Kit, developed by the SHRM, offers the following tips for communicating across language barriers:

- Be patient. If you expect too much, too quickly, you will likely confuse and frustrate the people involved.
- Don't get angry if people don't understand at first. Give them time to process the information in a way that they can comprehend.
- If people speak English poorly but understand it reasonably well, ask them to demonstrate their understanding through actions rather than words.
- Avoid humor. Your jokes will likely be misunderstood and may even be misinterpreted as a joke at the worker's expense.
- Don't assume that people are unintelligent simply because they don't understand your language. Many immigrants doing menial jobs in their adopted countries were doctors, lawyers, and educators in their native countries.
- Speak slowly and clearly, and avoid the tendency to yell.
- Use face-to-face communication whenever possible. Over-the-phone communication is often more difficult when a language barrier is involved.
- Use pictures or drawings to get your point across, if necessary.

3.3.0 Cultural Differences

As workers from a multitude of backgrounds and cultures are brought together, there are bound to be differences and conflicts in the workplace. These conflicts are often due to misunderstandings and lack of knowledge about each other's cultures.

To overcome cultural conflicts, the SHRM suggests the following:

- *Define the problem from both points of view* – How does each person involved view the conflict? What does each person think is wrong? This involves moving beyond traditional thought processes to consider alternate ways of thinking.
- *Uncover cultural interpretations* – What assumptions are being made based on cultural programming? By doing this, you as the supervisor may realize what motivated an employee to act in a particular manner. For instance, an employee may call in sick for an entire day to take his or her spouse to the doctor. In some cultures, such an action is based on responsibility and respect for family.
- *Find the strengths in the cultures that can make the team stronger* – Devise a solution that works for both parties involved. The purpose is to recognize and respect other's cultural values, and to work out mutually acceptable alternatives.

3.4.0 Sexual Harassment

In today's business world, men and women are working side-by-side in careers of all kinds. There are no longer male occupations or female jobs; therefore, men and women are expected to relate to each other in new ways that are uncharacteristic of the past.

As women make the transition into traditionally male industries, such as construction, the likelihood of sexual harassment increases. Sexual harassment is defined as unwelcome behavior of a sexual nature that makes someone feel uncomfortable in the workplace by focusing attention on their gender instead of on their professional qualifications. Sexual harassment can range from telling an offensive joke or hanging a poster of a swimsuit-clad man or woman in one's cubicle to making sexual comments or physical advances within a work environment.

Historically, sexual harassment was thought to be an act performed by men with power within an organization against women in subordinate positions. However, as the number of sexual harassment cases has shown over the years, this is not the only type of sexual harassment that happens.

Sexual harassment can occur in a variety of circumstances, including but not limited to the following:

- The victim as well as the harasser may be a woman or a man. The victim and the harasser do not have to be of opposite sex.
- The harasser can be the victim's supervisor, an agent of the employer, a supervisor in another area, a co-worker, or a non-employee.
- The victim does not have to be the person harassed but could be anyone affected by the offensive conduct.
- Unlawful sexual harassment may occur without economic injury to or firing of the victim.
- The harasser's conduct must be unwelcome.

The Equal Employment Opportunity Commission (EEOC) enforces sexual harassment laws within industries. When investigating allegations of sexual harassment, the EEOC looks at the whole record, including the circumstances and the context in which the alleged incidents occurred. A decision regarding the allegations is made from the facts on a case-by-case basis.

Prevention is the best tool for eliminating sexual harassment in the workplace. The EEOC encourages employers to take necessary steps to prevent sexual harassment from occurring. Employers should clearly communicate to employees that sexual harassment will not be tolerated. They can do so by developing a policy on sexual harassment, establishing an effective complaint or grievance process, and taking immediate and appropriate action when an employee complains.

3.5.0 Gender and Minority Discrimination

With the increase of women and minorities in the workforce, there is more room for gender and minority discrimination. Consequently, many business practices, including the way employees are treated, the organization's hiring and promotional practices, and the way people are compensated, are being analyzed for equity. More attention is being placed on fair recruitment, equal pay for equal work, and promotions for women and minorities in the workplace.

The EEOC requires that companies are equal opportunity employers. This means that organizations hire the best person for the job, without regard for race, sex, religion, age, etc. Once traditionally a male-dominated industry, construction companies are moving away from this notion and are actively recruiting and training women, younger workers, people from other cultures, and workers with disabilities to compensate for the shortage of skilled workers.

Despite the positive change in recruitment procedures, there are still inequities when it comes to pay and promotional activities associated with women and minorities. Individuals and federal agencies continue to win cases in which women and minorities have been discriminated against in the workplace, thus demonstrating that discrimination persists. The EEOC reported that between 1992 and 1997 two to three million dollars was awarded each year to those winning sexual discrimination cases. Between 1997 and 2006, the number of race-based charges of discrimination reported to the EEOC unfortunately held steady, at between 27,000 and 29,000 cases per year. The numbers are not much better for gender-based discrimination; between 1997 and 2006, the EEOC received an average of 24,000 complaints per year.

To prevent discrimination cases, employers should develop valid, job-related criteria for hiring, compensation, and promotion. These measures should be used consistently for every applicant interview, employee performance appraisal, etc. Therefore, all employees responsible for recruitment and selection, supervision, and evaluating job performance should be trained on how to use the job-related criteria legally and effectively.

4.0.0 ♦ CONSTRUCTION PROJECTS

As a supervisor, you will need to understand the wide range of projects in the construction industry, the categories of construction, and how each contributes to the needs of the overall industry. You will also need to be aware of how different crafts work together on a single project. The following sections show how work is organized and how it flows through different construction positions.

4.1.0 Major Categories of Construction Projects

The four major categories of construction are residential, commercial and institutional, industrial, and civil. In all four construction categories, there are several common threads that unite the industry:

- Construction helps satisfy the basic need for shelter and control of our environment.
- Special skills and professions are required.
- Materials and technology continue to improve.
- Building structures reflect the changing needs of society.

4.1.1 Residential Construction

Residential construction involves the construction of private homes. Unless the home is a very large one, it is unlikely to be a job involving pipefitters as such.

4.1.2 Commercial and Institutional Construction

Commercial and institutional construction includes building and modernizing stores, hospitals, schools, apartment houses, warehouses, restaurants, and office buildings. These projects range from small-scale jobs, involving only a minimum of time and costs, to large-scale jobs with contracts of several million dollars, which last many years.

Field personnel with many specialties are involved in commercial and industrial construction. Supervisors must be more involved in the activities than in residential construction. In addition, commercial and institutional construction projects usually involve the use of materials and systems not found in residential construction. Such materials include structural steel, pre-cast concrete, sprinkler systems, escalators, and elevators.

4.1.3 Industrial Construction

Industrial construction refers to building new manufacturing plants and modernizing and expanding existing plants. Because manufacturing itself is so diverse, industrial construction projects vary widely. They range from building an addition to a small electronics assembly plant to the construction of super-sized nuclear power plants.

4.1.4 Civil Construction

Civil construction refers to large projects needing civil engineering design. They are often public projects. Examples include the construction of roads, bridges, runways, utilities, and dams.

5.0.0 ♦ THE CONSTRUCTION ORGANIZATION

An organization is concerned with the relationships among the people within the company. As a supervisor, you need to be aware of two types of organizations, formal organizations and informal organizations.

A formal organization exists when the activities of the people within the work group are coordinated toward the attainment of a goal. An example of a formal organization is a work crew consisting of four carpenters and two laborers led by a supervisor all working together on a job.

A formal organization is typically documented on an organizational chart, which outlines all of the positions that make up an organization and how those positions are related. Some organizational charts even depict the people within each position and the person to whom they report, which is also known as span of control. An efficient span of control consists of a manager who directly supervises no more than six people.

An informal organization can be described as a group of individuals having a common goal, but for different reasons. An example of this would be a plane full of passengers flying to the same city, each with a different reason for doing so, and dispersing once the city is reached.

An informal organization allows for communication among its members so they can perform as a group. It also establishes patterns of behavior that help them to work as a group. A pattern may be a specific manner of dressing, speaking alike, or developing conformity.

Both types of organizations establish the foundation for how communications flow. The formal structure is the means used to delegate authority and responsibility and to exchange information. The informal structure is used to exchange information. The grapevine is an example of an informal organization.

Members in an organization perform best when each member achieves the following:

- Understands the job and how it will be done
- Communicates effectively with others in the group
- Knows his or her role in the group
- Knows who has the authority and responsibility

5.1.0 Division of Responsibility

The conduct of a construction business involves certain functions. In a small organization, responsibilities may be divided between one or two people. However, in a larger organization with many different and complex activities, responsibilities may be grouped into similar activity groups, and the responsibility for each group assigned to department managers. In either case, the following major departments exist in construction companies:

- *Executive* – This office represents top management. It is responsible for the success of the company through short-range and long-range planning.
- *Accounting* – This office is responsible for all record-keeping and financial transactions, including payroll, taxes, insurance, and audits.
- *Contract administration* – This office prepares and executes contractual documents with owners, subcontractors, and suppliers. Construction specialties such as pipefitting may have individual contractors. Usually, a construction contract will have a general contractor, who is responsible for all the trades involved.
- *Purchasing* – This office obtains material prices and then issues purchase orders. The purchasing office also obtains rental and leasing rates on equipment and tools.
- *Estimating* – This office is responsible for recording the quantity of material on the jobs, the takeoff, pricing labor and material, analyzing subcontractors' bids, and bidding on projects.
- *Construction operation* – This office plans, controls, and supervises all construction field activities.

Other divisions of responsibility a company may create involve architectural and engineering design functions. These divisions usually become separate departments.

5.2.0 Authority and Responsibility

As an organization grows, the manager must ask others to perform many duties so that he or she can concentrate on managing. Managers typically assign activities to their subordinates or workers through a process called delegation.

When delegating activities, the supervisor assigns others the responsibility to perform the designated tasks. Delegation implies responsibility, which refers to the obligation of an employee to perform the duties.

Along with responsibility comes authority. Authority is the power to act or make decisions in carrying out an assignment, such as hiring an applicant. The kind and amount of authority you will have as a supervisor or worker depends on the company for which you work.

Authority and responsibility must be balanced so employees can carry out their tasks. In addition, delegation of sufficient authority is needed to make an employee accountable to the supervisor for the results. Accountability is the act of holding an employee responsible for completing the assigned activities. Even though authority is delegated, the final responsibility always rests with the supervisor.

5.3.0 Job Descriptions

Many companies furnish each candidate a written job description, which explains the job in detail. Job descriptions set a standard for the employee, make judging performance easier, clarify the tasks each person should handle, and simplify the training of new employees.

Each new employee should understand all the duties and responsibilities of the job after reviewing the job description. Thus, the time it takes for the employee to make the transition from being a new and uninformed employee to a more experienced member of a production team is shortened.

A job description need not be long, but it should be detailed enough to ensure there is no misunderstanding of the duties and responsibilities of the position. The job description should contain all the information necessary to evaluate the employee's performance.

A job description should contain the following:
- Job title
- General description of the position
- Specific duties and responsibilities
- The supervisor to whom the position reports
- Other requirements, such as required tools/equipment for the job

A sample job description is shown in *Figure 2*.

5.4.0 Policies and Procedures

Most construction companies have formal policies and procedures established to help the supervisor carry out his or her duties. A policy is a general statement establishing guidelines for a specific activity. Examples include policies on vacations, rest breaks, workplace safety, or checking out power hand tools.

Procedures are the ways that policies are carried out. For example, a procedure written to implement a policy on workplace safety would include guidelines for reporting accidents and general safety procedures that all employees are expected to follow.

A sample of a policy and procedure is shown in *Figure 3*.

6.0.0 ◆ BECOMING A LEADER

The crew leader is generally selected and promoted from a construction work crew. If you are selected for crew leader, that selection will frequently be based on your ability to accomplish tasks, to get along with others, to meet schedules, and to stay within the budget. You must lead the team to work safely and provide a quality product.

Making the transition from a craftworker to a crew leader can be difficult, especially when you find yourself in charge of supervising a group of peers. Crew leaders are no longer responsible for their work alone; rather, they are accountable for the work of an entire crew of people with varying skill levels and abilities, a multitude of personalities and work styles, and different cultural and educational backgrounds. Supervisors must learn to put their personal relationships aside and work for the common goals of the team.

As you move from a craftworker position to the role of a crew leader, you will find that more hours will be spent supervising the work of others than actually performing the technical skill for which you have been trained. *Figure 4* represents the percentage of time craftworkers, crew leaders, superintendents, and project managers spend on technical and supervisory work as their management responsibilities increase.

Your success as a new crew leader is directly related to your ability to make the transition from crew member into a leadership role. Leadership includes possessing the traits and skills to motivate others to follow and perform.

6.1.0 Characteristics of Leaders

Leadership traits are similar to the skills that a supervisor needs to be effective. Although the characteristics of leadership are varied, there are some definite commonalities among effective leaders.

First and foremost, effective leaders lead by example. In other words, they work and live by the standards that they establish for their crew members or followers, making sure that they practice what they preach.

Second, effective leaders tend to have a high level of drive, determination, persistence, or a

---

Position:
Front-line Construction Supervisor – Field

General Summary:
First line of supervision on a construction crew installing underground and underwater cathodic protection systems.

Reports To:
Field Superintendent

Duties and Responsibilities:
- Oversee crew
- Provide instruction and training in construction tasks as needed
- Make sure proper materials and tools are on the site to accomplish tasks
- Keep project on schedule
- Enforce safety regulations

Knowledge, Skills, and Experience Required:
- Extensive travel throughout the Eastern United States, home base in Atlanta
- Ability to operate a backhoe and trencher
- Valid commercial drivers license with no DUI violations
- Ability to work under deadlines with the knowledge and ability to foresee problem areas and develop a plan of action to solve the situation

Figure 2 ◆ Job description.

> **WORKPLACE SAFETY POLICY:**
>
> Your safety is the constant concern of this company. Every precaution has been taken to provide a safe workplace.
>
> Common sense and personal interest in safety are still the greatest guarantees of your safety at work, on the road, and at home. We take your safety seriously and any willful or habitual violation of safety rules will be considered cause for dismissal. We are sincerely concerned for the health and well being of each member of the team.
>
> **Workplace Safety Procedure:**
> To ensure your safety, and that of your co-workers, please observe and obey the following rules and guidelines:
>
> - Observe and practice the safety procedures established for the job.
> - In case of sickness or injury, no matter how slight, report at once to your supervisor. In no case should an employee treat his own or someone else's injuries or attempt to remove foreign particles from the eye.
> - In case of injury resulting in possible fracture to legs, back, or neck, any accident resulting in an unconscious condition, or a severe head injury, the employee is not to be moved until medical attention has been given by authorized personnel.
> - Do not wear loose clothing or jewelry around machinery. It may catch on moving equipment and cause a serious injury.
> - Never distract the attention of another employee, as you might cause him or her to be injured. If necessary to get the attention of another employee, wait until it can be done safely.
> - Where required, you must wear protective equipment, such as goggles, safety glasses, masks, gloves, hairnets, etc.
> - Safety equipment such as restraints, pull backs, and two-hand devices are designed for your protection. Be sure such equipment is adjusted for you.
> - Pile materials, skids, bins, boxes, or other equipment so as not to block aisles, exits, fire-fighting equipment, electric lighting or power panel, valves, etc. FIRE DOORS AND AISLES MUST BE KEPT CLEAR.
> - Keep your work area clean.
> - Use compressed air only for the job for which it is intended. Do not clean your clothes with it and do not play with it.

Figure 3 ♦ Safety policy and procedure.

stick-to-it attitude. In the face of a challenging situation, leaders seek involvement and work through adversity to achieve their goal. When faced with obstacles, effective leaders don't get discouraged; instead, they identify the potential problems, make plans to overcome them, and work toward achieving the intended goal. In the event of failure, effective leaders learn from their mistakes and apply that knowledge to their future leadership attempts.

Third, effective leaders are effective communicators. Accomplishing this may require that the leader overcome issues such as language barriers, gender biases, or differences in personalities to ensure that each follower or member of the crew understands the established goals of the project.

Effective leaders have the ability to motivate their followers or crew members to work to their full potential and become effective members of the team. They try to develop their crew members' skills and encourage them to improve and learn as a means to contribute more to the team effort. Effective leaders tend to demand 100 percent from themselves and their teams, so they work hard to provide the skills and leadership necessary to do so.

Finally, effective leaders are organized planners. They know what needs to be accomplished,

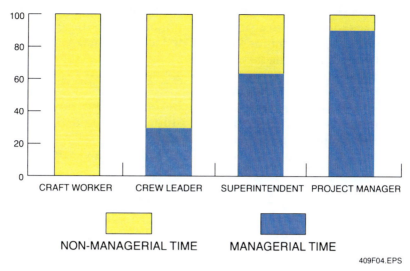

Figure 4 ♦ Percentage of time spent on technical and supervisory work.

and they use their resources to make it happen. Because they cannot do it alone, leaders enlist the help of their team members to share in the workload. Effective leaders delegate work to their crew members, and they implement company policies and procedures to ensure that the work is completed effectively, efficiently, and safely. When delegating work, make sure that the crew member understands what to do and the level of responsibility.

Finally, effective leaders have the self-confidence that allows them to make decisions and solve problems. If they are going to accomplish their goals, leaders must learn to take risks when appropriate. Leaders must be able to absorb information, assess courses of action, understand the risks, make decisions, and assume the responsibility for them. Some other major characteristics of leaders include the following:

- The ability to plan and organize
- Loyalty
- Fairness
- Enthusiasm
- The ability to teach and learn from others
- Initiative
- Salesmanship
- Good communication skills

6.2.0 Leadership Behavior

Followers have expectations of their leaders. Employees need to feel that their skills and abilities are valued and make a difference. In other words, they crave a feeling of job importance. They also look to their leaders to do the following:

- Set the example
- Suggest and direct
- Influence their actions
- Communicate effectively
- Make decisions and assume responsibility
- Be a loyal member of the group
- Abide by company policies and procedures

6.3.0 Functions of a Leader

The functions of a leader will vary with the environment, the group being led, and the tasks to be performed. However, there are certain functions common to all situations that the leader will be called upon to perform. Some of the major functions are as follows:

- Organize, plan, staff, direct, and control.
- Empower group members to make decisions and take responsibility for their work.
- Maintain a cohesive group by resolving tensions and differences among its members and between the group and those outside the group.
- Ensure that all group members understand and abide by company policies and procedures.
- Accept responsibility for the success and failures of the group's performance.
- Represent the group to others.
- Be sensitive to the differences of a diverse workforce.

6.4.0 Leadership Styles

There are three broad categories of leadership style that supervisors can adopt. At one extreme is the autocratic or dictator style of leadership, where the supervisor makes all of the decisions independently, without seeking the opinions of crew members. At the other extreme is the hands-off leadership style, where the supervisor lets the employees make all of the decisions. In the middle is the democratic style, where the supervisor seeks crew members' opinions and makes the appropriate decisions based on their input.

The following are some characteristics of each of the three leadership styles:

Autocratic leaders:
- Expect their crew members to work
- Seldom seek advice from their crew members
- Insist on solving problems alone
- Seldom permit crew members to assist each other
- Praise and criticize on a personal basis
- Have no sincere interest in creatively improving methods of operation or production

Hands-off leaders:
- Believe no supervision is best
- Rarely give orders
- Worry about whether they are liked by their crew members

Democratic leaders:
- Discuss problems with their crew members
- Listen
- Explain and instruct
- Give crew members a feeling of accomplishment by commending them when they do a job well
- Are friendly and available to discuss personal and job-related problems

Effective leadership takes many forms. The correct style for a particular situation or operation depends on the nature of the crew as well as the work it has to accomplish. For example, if the crew does not have enough experience for the job ahead, then a direct, autocratic style may be appropriate. The autocratic style of leadership is also effective when jobs involve repetitive operations that require little decision-making.

However, if a worker's attitude is an issue, then democratic action is required. In this case, providing the missing motivational factors may increase performance and result in the improvement of the worker's attitude. The democratic style of leadership is also useful when the work is of a creative nature because brainstorming and exchanging ideas with the crew members can be beneficial.

Hands-off leadership is not a very effective style. In these cases, either the leadership becomes the responsibility of a crew member who takes the initiative to assume the role, or chaos results due to lack of formal leadership from the supervisor.

6.4.1 Selecting a Leadership Style

In selecting the most appropriate style of leadership, you should consider the power base from which to work. There are three elements that make up a power base:

1. Authority
2. Expertise, experience, and knowledge
3. Respect, attitude, and personality

First, the company must give you, the supervisor, enough authority to do the job. This authority must be commensurate with responsibility, and it must be made known to crew members when they are hired so that they understand who is in charge.

Next, you must have an expert knowledge of the activities to be supervised in order to be effective. This is important because the crew members need to know that they have someone to turn to when they have a question or a problem, when they need some guidance, or when modifications or changes are needed.

Finally, respect is probably the most critical element of power. This derives from being fair to employees, by listening to their complaints and suggestions, and by utilizing incentives and rewards appropriately to motivate crew members. In addition, supervisors who have a positive attitude and a favorable personality tend to gain the respect of their crew members as well as their peers. Along with respect comes a positive attitude from the crew members.

6.5.0 Ethics in Leadership

The construction industry demands the highest standards of ethical conduct. Every day the crew leader has to make decisions which may have ethical implications. If you make an unethical decision as a supervisor, it hurts you, other workers, peers, and the company for which you work.

There are three basic types of ethics:

1. Business, or legal
2. Professional, or balanced
3. Situational

Business, or legal, ethics means adhering to all laws and regulations related to the issue. Professional, or balanced, ethics is carrying out all activities in such a manner as to be fair to everyone concerned. Situational ethics pertains to specific activities or events that may initially appear to be a gray area. For example, you may ask yourself,

"How will I feel about myself if my actions are published in the newspaper or if I have to justify my actions to my family, friends, and colleagues?"

Consider the example of professional football. The National Football League has rules by which the game is played. If someone breaks one of the rules, he is acting in an unethical manner from a business or legal sense. It is an accepted standard among professional football players that when they tackle another player it is to take him out of the play and not to purposely injure him. If one were to tackle someone with the motivation of causing an injury, this would be considered unethical.

If you were running down the sideline after catching a pass to score a touchdown, and in your attempt slightly stepped out of bounds (without anyone calling the ball dead at that point), and you did not voluntarily acknowledge that you did, you would be unethical from a situational standpoint. How would you feel the next day if a newspaper printed a picture showing your foot out of bounds and insinuating that your character is questionable because you did not acknowledge breaking the rules?

There are going to be many times when, as a supervisor, you will be put into a situation where you will need to assess the ethical consequences of an impending decision. For instance, should a supervisor continue to keep one of his or her crew working who has broken into a cold sweat due to overheated working conditions just because the superintendent says the activity is behind schedule? Or should a supervisor, who is the only one aware that the reinforcing steel placed by his or her crew was done incorrectly, correct the situation before the concrete is placed in the form? If a crew leader is ever asked to carry through on an unethical decision, it is up to him or her to inform the supervisor of the unethical nature of the issue, and if still requested to follow through, refuse to act.

7.0.0 ♦ PROBLEM SOLVING AND DECISION MAKING

Like it or not, problem solving and decision making are a large part of every supervisor's daily work. The supervisor's first job is to get the job done on time, within budget, and safely. There will always be problems to be resolved and decisions to be made, especially in fast-paced, deadline-oriented industries such as construction. Sometimes, the difference between problem solving and decision making is not clear. Decision making refers to the process of choosing an alternative course of action in a manner appropriate for the situation. Problem solving involves determining the difference between the way things are and the way things should be, and finding out how to bring the two together. The two activities are interrelated because in order to make a decision, you may also have to use problem-solving techniques.

7.1.0 Types of Decisions

Some decisions are routine or simple. These types of decisions can be made based on past experiences. An example would be deciding how to get to and from work. If you've worked at the same place for a long time, you are already aware of the options for traveling to and from work (take the bus, drive a car, carpool with a co-worker, take a taxi, etc.). Based on past experiences with the options identified, you can make a decision of how best to get to and from work.

On the other hand, some decisions are non-routine or more difficult. These types of decisions require more careful thinking about how to carry out an activity by using a formal problem-solving technique. An example is planning a trip to a new vacation spot. If you are not sure how to get there, where to stay, what to see, etc., one option is to research the area to determine the possible routes, hotel accommodations, and attractions. Then you will have to make a decision about which route to take, what hotel to choose, and what sites to visit, without the benefit of direct past experience with the options.

7.2.0 Formal Problem-Solving Techniques

To make non-routine decisions, the following procedure can be used as a part of formal problem solving.

Step 1   Recognize that a problem exists.

A problem is the difference between the way things are and the way that you want them to be. Solving a problem refers to eliminating the differences between what's existing and what's desired, or making things turn out the way you want them.

Consider the following example:

It is 3:30 p.m. Friday afternoon, and you are walking back to the job site trailer after making the afternoon rounds on your job. As you pass the carpenter's workshop, you can't help but notice a lot of good looking, useable lumber in the scrap pile. This bothers you, so you proceed to ask the carpenter in charge why so much good lumber is in the scrap pile. Mary, the head carpenter, states she was not aware that there was so much useable lumber being scrapped.

You both decide to take a closer look at the scrap pile to see just how much is salvageable. You both conclude that a lot of usable lumber is being thrown away. Upon further questioning of the carpenter, you find that she has been sick the last two days, and she left her helper, Tom, to do the work.

Investigating even further, you find that Tom just graduated from high school and is enrolled in the first semester of the first-year carpentry program. According to his instructor, Tom appears willing to learn and performs satisfactorily in the classroom. Only the basics of carpentry have been taught to date (math, trade hand tools, and print reading).

Rereading the scenario, the first thing to do to solve the problem is to identify the signals (statements) that indicate that a problem exists. In this case, the main signal is that "a lot of good looking, useable lumber is in the scrap pile."

There may be other areas related to the one noted above. If so, what are they? As the reader can see, the signal is the thing that will motivate the supervisor to find a solution to the problem.

Step 2 Determine what you know or assume you know about the problem, and separate facts from non-facts.

This second step involves answering questions about the problems. Specifically, the *what, when, where,* and *who* should be determined. Once these questions have been considered, the information must be categorized as factual or non-factual.

Referring to the example above, some responses to an analysis of the problem may be as follows:

What?
- There is too much good, usable lumber in the scrap pile.
- There is too much waste of materials on this job site.
- The job has gone over budget because of this waste.

When?
- The problem was discovered on Friday afternoon.

Where?
- The problem is the scrap pile near the carpenter's workshop.

Who?
- Mary, the carpenter in charge, was off sick for two days, and she left a helper in charge.
- The carpenter acted incorrectly by leaving the helper in charge.
- The helper, Tom, doesn't know how to do his job.

Based on the analysis above, some of the statements are facts while the others are assumptions, judgments, expressions of frustration, and/or blame. To solve the real problem, you must eliminate the non-factual items and deal only with the facts.

Facts:
- There is too much good, usable lumber in the scrap pile.
- The problem was discovered on Friday afternoon.
- The problem is in the scrap pile near the carpenter's workshop.
- The carpenter in charge was off sick for two days, and she left a helper in charge.

Non-facts:
- There is too much waste of materials on this job site.
- The job has gone over budget because of this waste.
- The carpenter acted incorrectly by leaving the helper in charge.
- The helper doesn't know how to do his job.

Step 3 State the problem.

Using only facts, formulate problem statements. A fact is a statement of the problem when it meets the following criteria:

- A fact notes the difference between "what is" and "what should be."
- A fact indicates that the present situation has potential for change.

Using the factual statements from the example case, the problem statements are as follows:

- There is much good lumber in the scrap pile that could be used in the construction of the job.
- The carpenter's helper may not know how to cut lumber to minimize waste.

Step 4 Develop objectives that will eliminate the problem.

Describing the desired end result (an objective) is a turning point in the problem-solving sequence. At this point, decision making comes into play. It involves choosing to do one of two things: to act on only one idea that comes into mind or to act on what appears to be the most effective of several alternatives.

Written or stated objectives are the basis of the decision-making process. The more precise the objective, the easier it will be to choose the most effective and efficient plan of action.

Continuing with the example above, some of your objectives may be the following:

- By the end of business on Monday, I will have found out whether or not Tom knows how to cut lumber in a way that will reduce waste.
- By the end of business on Tuesday, I will have had a discussion with Mary to ensure that she realizes that training helpers is a part of her job.
- By the end of business on Wednesday, Mary will have prepared an on-the-job training program for Tom.
- By the end of business on Friday, Tom will have begun an on-the-job training program on how to select and cut lumber efficiently.

Step 5 Develop alternate solutions, and select one that will solve the problem.

At this point, you should list all possible actions that can be taken to accomplish the objectives, thereby solving the problem. When developing the list, judgments should not be made about each alternative. Instead, each possible action should be listed, and then you should weigh the alternatives to make a decision.

The previous example might have the following list of alternatives:

- Talk to Tom to determine whether or not he is aware of how to cut lumber efficiently; if not, provide training to enable him to do so.
- Provide an on-the-job training program.
- Discuss the problem with Mary and determine the best solution together.
- Inform Mary that it's her responsibility as a lead person to provide training where and whenever needed. Therefore, instruct her to develop and implement a training program.
- Follow up after the program to see if, in fact, Tom can now perform satisfactorily.

Step 6 Develop a plan of action.

Once the solution to the problem has been determined, the next step is to decide who is responsible for carrying out the plan and the deadline for accomplishing it, if not already stated in the objective(s).

For this example, one definite plan of action would be:

1. Meet with Mary to discuss the problem. Get a commitment from her that by Monday she will have discussed the problem with her helper and will report back with the outcome.
2. Upon meeting with Mary the second time, obtain a commitment from her to develop a week-long, on-the-job training program for Tom. Inform Mary that it will be her duty to implement the training program and follow up on it. The plan is due by Tuesday.
3. Meet with Mary to review and make any needed revisions to her plan.
4. Follow up next week to ensure that Tom has successfully completed the training program.

Once all six steps of the formal problem-solving technique are completed, you must follow up to ensure that the action plan was completed as intended. If not, corrective measures should be taken to see that the plan is carried out. In addition, you should take steps to prevent the same problem from occurring again in the future.

7.3.0 Dealing with Leadership Problems

When you are responsible for leading others, it is inevitable that you will encounter problems and be forced to make decision about how to respond. Some problems will be relatively simple to handle, such as covering for a sick crew member. Other problems will be complex and much more difficult. Common problems include poor attitudes toward the workplace, an employee's inability to work with others, and absenteeism and turnover.

7.3.1 Poor Attitude Towards the Workplace

Sometimes, employees have poor attitudes towards the workplace because of bad relationships with their fellow employees, negative outlooks on their supervision, or a dislike of the job in general. Whatever the case, it is important that in your role as supervisor you determine the cause of the poor attitude.

The best way to determine the cause of a poor attitude is to talk with that employee one-on-one, listening to what the employee has to say and asking questions to uncover information. Once this conversation has occurred and the facts have been assessed, you can determine how to correct the situation and turn the negative attitude into a positive one.

If you discover that the problem stems from factors in the workplace or the surrounding environment, you have several choices. First, you can move the worker from the situation to a more acceptable work environment. Next, you can change that part of the work environment found to be causing the poor attitude. Finally, you can take steps to change the employee's attitude so that the work environment is no longer a negative factor.

7.3.2 Inability to Work with Others

Sometimes you will encounter situations where an employee has a difficult time working with others on the crew. This could be a result of personality differences, an inability to communicate, or some other cause. Whatever the reason, you must address the issue and get the crew working as a team.

The best way to determine the reason for why individuals don't get along or work well together is to talk to the parties involved. You should speak openly with the employee as well as the other individual(s) to find out why.

Once the reason for the conflict is found, you can determine how to respond. There may be a way to resolve the problem and get the workers communicating and working as a team again. On the other hand, there may be nothing that can be done that will lead to a harmonious situation. In this case, you would either have to transfer the employee to another crew or have the problem crew member terminated. This latter option should be used as a last measure and should be discussed with your superiors or Human Resources Department.

7.3.3 Absenteeism and Turnover

Absenteeism and turnover are big problems on construction jobs. Without workers available to do the work, jobs are delayed, and money is lost.

Absenteeism refers to workers missing their scheduled work time on a job. Absenteeism has many causes, some of which are inevitable. For instance, people get sick, they have to take time off for family emergencies, and they have to attend family events such as funerals. However, there are some causes of absenteeism that could be prevented by the supervisor.

The most effective way to control absenteeism is to make the company's policy clear to all employees. Companies that do this find that chronic absenteeism diminishes as a problem. New employees should have the policy explained to them. This explanation should include the number of absences allowed and the reasons for which sick or personal days can be taken. In addition, all workers should know how to inform their supervisors when they miss work and understand the consequences of exceeding the number of sick or personal days allowed.

Once the policy on absenteeism is explained to employees, you must be sure to implement it consistently and fairly. If the policy is administered equally, employees will likely follow it. However, if the policy is not administered equally and some employees are given exceptions, then it will not be effective. Consequently, the rate of absenteeism will increase.

Despite having a policy on absenteeism, there will always be employees who are chronically late or miss work. In cases where an employee abuses the absenteeism policy, discuss the situation directly with the employee. Confirm that the employee understands the company's policy and insist that the employee comply with it. If the employee's behavior continues, disciplinary action may be in order.

Turnover refers to the loss of an employee that is initiated by that employee. In other words, the employee quits and leaves the company to work elsewhere or is fired for cause.

Like absenteeism, there are some causes of turnover that cannot be prevented and others that can. For instance, it is unlikely that a supervisor could keep an employee who finds a job elsewhere earning twice as much money. However, you can prevent some employee turnover situations. You can work to ensure safe working conditions for your crew, treat workers fairly and consistently, and help promote good working conditions. The key to doing so is communication. You need to know the problems if you are going to be able to successfully resolve them.

Some of the major causes of turnover include the following:

- *Uncompetitive wages and benefits* — Workers may leave one construction company to go to another that pays higher wages and/or offers better benefits. They may also leave to go to another industry that pays more.
- *Lack of job security* — Workers leave to find more permanent employment.
- *Unsafe project sites* — Workers leave for safer projects.
- *Unfair/inconsistent treatment by their immediate supervisor*
- *Poor working conditions*

Essentially, the same actions described above for absenteeism are also effective for reducing turnover. Past studies have shown that maintaining harmonious relationships on the job site will go a long way in reducing both turnover and absenteeism. This will take effective leadership on the part of the supervisor.

8.0.0 ♦ SAFETY RESPONSIBILITIES

The construction industry loses millions of dollars every year because of work-site accidents. Work-related injuries, sickness, and deaths have caused untold human misery and suffering. Project delays and budget overruns from construction injuries and fatalities are extremely costly. Work-site accidents also erode the overall morale of the crew.

Despite the many hazards associated with construction, experts believe that applying preventive safety measures will drastically reduce the number of accidents. Each employer must set up a safety and health program to manage workplace safety and health and to reduce injury, illness, and fatalities. The program must be appropriate to the conditions of the workplace. It should consider the number of workers employed and the hazards to which they are exposed while at work.

To be successful, the safety and health program must have management leadership and employee participation. The main way that supervisors to encourage safety is by example. If you always perform every task with safe procedures, employees will understand that safety is important to you. In addition, training and informational meetings play an important part in effective programs. As a crew leader or supervisor, you are responsible for consistently and effectively carrying out and enforcing the company's safety program and making sure all workers perform their tasks safely. To accomplish this, you must satisfy the following requirements:

- Be aware of the cost of accidents.
- Understand all federal, state, and local government safety regulations and penalties for violation.
- Be involved in training workers in safe-work methods.
- Conduct safety meetings.
- Get involved in safety inspections, accident investigations, and fire protection and prevention.
- Attain the required safety certifications appropriate to your position.
- Keep proper safety records.
- Investigate accidents according to proper procedures.

8.1.0 Safety Information and Training

Employers and supervisors should provide periodic information and training to new and long-term employees. If you are a supervisor, implement safety training as often as necessary, especially when safety and health information changes or workplace conditions create new hazards.

You must be sure that workers are capable of doing the work in a safe manner. You can accomplish this by providing information one-on-one or to a group. You should do the following to keep workers safe:

- Define the task.
- Explain how to do the task safely.
- Explain which tools and equipment to use and how to use them safely.
- Identify the necessary personal protective equipment (PPE).
- Explain the nature of the hazards in the work and how to recognize them.
- Stress the importance of personal safety and the safety of others.
- Hold regular safety meetings with the crew's input.
- Review material safety data sheets (MSDS) as necessary.

Review Questions

1. The construction industry should provide training for craftworkers and supervisors _____.
   a. to ensure that there are enough future workers
   b. since growth demands are anticipated to be exceeded by replacement needs
   c. in order to update the skills of older workers who are retiring at a later age than they previously did
   d. even though younger workers are now less likely to seek jobs in other areas than they were 10 years ago

2. Construction companies traditionally offer craftworker training _____.
   a. that a supervisor leads in a classroom setting
   b. that a craftworker leads in a classroom setting
   c. in a hands-on setting, where craftworkers learn from a co-worker or supervisor
   d. on a self-study basis to allow craftworkers to proceed at their own pace

3. One way to prevent sexual harassment in the workplace is to _____.
   a. require employee training in which the potentially offensive subject of stereotypes is carefully avoided
   b. develop a consistent policy with appropriate consequences for engaging in sexual harassment
   c. communicate to workers that the victim of sexual harassment is the one who is being directly harassed, not those affected in a more indirect way
   d. educate workers to recognize sexual harassment for what it is—unwelcome conduct by the opposite sex

4. Employers can minimize all types of workplace discrimination by doing all of the following *except* _____.
   a. investigating those complaints that are serious
   b. looking for differences between the number in the workforce and number of qualified workers
   c. hiring based on a consistent list of job-related requirements
   d. providing sexual harassment training

5. One of the four major categories of construction projects is _____.
   a. civil, which involves such projects as road and bridge construction
   b. residential, which involves any sort of dwelling such as homes as well as apartments
   c. commercial and institutional, which use materials similar to those in residential construction but on a larger scale
   d. institutional and industrial, which involves school, offices, and factories

6. A formal organization uses an organizational chart to _____.
   a. depict all companies with whom it conducts business
   b. show all customers with whom it conducts business
   c. show the relationships among the existing positions in the generic sense, typically excluding the actual people within the positions
   d. show the relationships among the existing positions; may even include the actual people in the positions

Review Questions

7. Authority is defined as _____.
   a. making decisions without ever having to discuss things with others
   b. giving an employee a particular task to perform
   c. making an employee responsible for the completion and results of a particular duty
   d. having the power to perform an action or make decisions, such as promoting someone

8. Job descriptions are used in order to _____.
   a. specify to job candidates whether a company uses a formal or informal organizational structure
   b. give details to potential employees about how to perform specific duties
   c. give information regarding the responsibilities of the potential job candidate
   d. inform the potential candidate about the company's benefits

9. The purpose of a policy is to _____.
   a. implement company guidelines regarding a particular activity
   b. specify what tools and equipment are required for a job
   c. list all information necessary to judge an employee's performance
   d. inform employees about the future plans of the company

10. Of the three styles of leadership, the _____ style would be effective in dealing with a craftworker's negative attitude.
    a. hands-off, followed by autocratic
    b. autocratic, followed by hands-off
    c. democratic
    d. autocratic

Summary

Making the professional advancement from craftworker to crew leader or supervisor requires you to master further skills in management, communications, ethics, and leadership. Knowledge of your trade will not be as important as your ability to manage and lead a diverse work group into performing at the level of their potential. To do this, you must promote a fair work environment by being able to assemble workers from various cultural backgrounds, prevent discrimination and harassment, and consistently enforce policies and procedures. You must also promote a safe working environment by implementing and enforcing safety training as necessary. Your team will look to you for leadership and direction, so you must exhibit the qualities of a strong, democratic leader, and be able to make sound decisions and identify and solve problems on the job. Understanding how to make workers accountable to you and creating a safe and fair work environment will reduce attitude problems, maintain good morale among your crew, and improve the quality and efficiency of your crew's work.

Notes

Resources & Acknowledgments

Additional Resources

This module is intended to present thorough resources for task training. The following references are suggested for further study. These are optional materials for continued education rather than for task training.

American Medical Association (AMA), www.ama-assn.org

American Society for Training and Development (ASTD), www.astd.org

Architecture, Engineering, and Construction Industry (AEC), www.aecinfo.com

Bureau of Labor Statistics, www.bls.gov

CIT Group, www.citgroup.com

Contren® Management Learning Series, www.nccer.org

Equal Employment Opportunity Commission (EEOC), www.eeoc.gov

Jamestown Area Labor Management Committee (JALMC), www.jalmc.com

Knowledge Center's Manager's Toolkit, www.knowledecenters.versaware.com

National Association of Women in Construction (NAWIC), www.nawic.org

National Center for Construction Education and Research (NCCER), www.nccer.org

National Institute of Occupational Safety and Health (NIOSH), www.cdc.gov/niosh

National Safety Council, www.nsc.org

Occupational Safety and Health Administration, www.osha.gov

Society for Human Resources Management (SHRM), www.shrm.org

United States Census Bureau, www.census.gov

United States Department of Labor, www.dol.gov

NCCER CURRICULA — USER UPDATE

NCCER makes every effort to keep its textbooks up-to-date and free of technical errors. We appreciate your help in this process. If you find an error, a typographical mistake, or an inaccuracy in NCCER's curricula, please fill out this form (or a photocopy), or complete the online form at **www.nccer.org/olf**. Be sure to include the exact module ID number, page number, a detailed description, and your recommended correction. Your input will be brought to the attention of the Authoring Team. Thank you for your assistance.

Instructors – If you have an idea for improving this textbook, or have found that additional materials were necessary to teach this module effectively, please let us know so that we may present your suggestions to the Authoring Team.

NCCER Product Development and Revision
13614 Progress Blvd., Alachua, FL 32615
Email: curriculum@nccer.org
Online: www.nccer.org/olf

❑ Trainee Guide ❑ AIG ❑ Exam ❑ PowerPoints Other _____

Craft / Level: _____ Copyright Date: _____

Module ID Number / Title: _____

Section Number(s): _____

Description: _____

_____

_____

_____

Recommended Correction: _____

_____

_____

_____

Your Name: _____

Address: _____

Email: _____ Phone: _____

Glossary of Trade Terms

Acetylene: A gas composed of two parts carbon and two parts hydrogen, commonly used in combination with oxygen to cut, weld, and braze steel.

Alloy: A metal that has had other elements added to it, which substantially changes its mechanical properties.

Base line: A straight line drawn around a pipe to be used as a measuring point.

Body: The main part of the valve. It contains the disc, seat, and valve ports. The body of the valve is directly connected to the piping by threaded, welded, mechanically joined, or flanged ends.

Bonnet: The part of the valve that contains the trim. The bonnet is located above the body.

Branch: A line that intersects with another line.

Carbon: An element which, when combined with iron, forms various kinds of steel. In steel, it is the carbon content that affects the physical properties of the steel.

Chord: A straight line crossing a circle that does not pass through the center of the circle.

Compression coupling: A mechanical fitting that is compressed onto a pipe, tube, or hose.

Conductivity: A measure of the ability of a material to transmit electron flow.

Condensate: The liquid byproduct of cooling steam.

Constituents: The elements and compounds, such as metal oxides, that make up a mixture or alloy.

Critical temperature: The temperature at which iron crystals in a ferrous-based metal transform from being face-centered to body-centered. This dramatically changes the strength, hardness, and ductility of the metal.

Cutback: The point at which a miter fitting is to be cut.

Differential pressure: The measurement of one pressure with respect to another pressure, or the difference between two pressures.

Dimension: A measurement between two points on a drawing.

Direct pressure: Flow pushing the sealing element of the valve into the seat and improving closure.

Disc: The moving part of a valve that directly affects the flow through the valve.

Drip leg: A drain for condensate in a steam line placed at a low point in the line and used with a steam trap.

Ferrule: A bushing placed over a tube to tighten it.

Fishmouth: A fabricated 90-degree intersection of pipe.

Flaring: Increasing the diameter at the end of a pipe or tube.

Flash steam: Steam formed when hot condensate is released to lower pressure and re-evaporated.

Full port: The maximum internal opening for flow through a valve that matches the ID of the pipe used.

Hardenability: A characteristic of a metal that makes it able to become hard, usually through heat treatment.

Head clearance: The amount of space needed to install a hot tap machine on a pipe.

Heat-affected zone (HAZ): The area of the base metal that is not melted but whose mechanical properties have been altered by the heat of the welding.

Hot tap: To make a safe entry into a pipe or vessel operating at a pressure or vacuum under controlled conditions without losing product.

Induction heating: The heating of a conducting material by means of circulating electrical currents induced by an externally applied alternating magnetic field.

Interpass heat: The temperature to which a metal is heated while an operation is performed on the metal.

Line stop: A device used to temporarily contain the flow or pressure of a product inside a pipe while a branch connection is being made.

Nonferrous metal: A metal that contains no iron and is therefore nonmagnetic.

Ordinate lines: Straight lines drawn along the length of the pipe connecting the ordinate marks.

Ordinate: A division of segments obtained by dividing the circumference of a pipe into equal parts.

Oxidation: A chemical reaction that increases the oxygen content of a compound.

Packing: The material between the valve stem and bonnet that provides a leakproof seal and prevents material from leaking up around the valve stem.

Port: The internal opening for flow through a valve.

Glossary of Trade Terms

Postheat: The temperature to which a metal is heated after an operation is performed on the metal.

Preheat weld treatment (PHWT): The controlled heating of the base metal immediately before welding begins.

Preheat: The temperature to which a metal is heated before an operation is performed on the metal.

Quenching: Rapid cooling of a hot metal using air, water, or oil.

Reamer: A tool used to enlarge, shape, smooth, or otherwise finish a hole.

Reverse pressure: Flow pushing the closure element out of the seat.

Revision: A change in a part on an engineering drawing that is noted on the drawing.

Saturated steam: Pure steam without water droplets that is at the boiling temperature of water for the given pressure.

Seat: The nonmoving part of the valve on which the disc rests to form a seal and close off the valve.

Segment: A part of a circle that is defined by a chord and the curve of the circumference.

Solder: An alloy, such as of zinc and copper, or of tin and lead, that, when melted, is used to join metallic surfaces.

Soldering flux: A chemical substance that aids the flow of solder. Flux removes and prevents the formation of oxides on the pieces to be joined by soldering.

Stem: The part of the valve that connects the disc to the valve operator. A stem can have linear, rotary, or helical movement.

Stress relieving: The even heating of a structure to a temperature below the critical temperature followed by a slow, even cooling.

Stress: The load imposed on an object.

Superheated steam: Saturated steam to which heat has been added to raise the working temperature.

Tempering: Increases the toughness of quenched steel and helps avoid breakage and failure of heat-treated steel. Also called *drawing*.

Title block: A section of an engineering drawing blocked off for pertinent information, such as the title, drawing number, date, scale, material, draftsperson, and tolerances.

Tracer: A steam line piped beside product piping to keep the product warm or prevent it from freezing.

Travel distance: The distance that the cutting bit moves from the top of the tapping valve to the pipe to be cut.

Trim: The internal parts of a valve that receive the most wear and can be replaced. The trim includes the stem disc, seat ring, disc holder or guide, wedge, and bushings.

Water hammer: A condition that occurs when hot steam comes in contact with cooled condensate, builds pressure, and pushes the water through the line at high speeds to slam into valves and other equipment with devastating effect.

Wetting: Spreading liquid filler metal or flux on a solid base metal.

Pipefitting Level Four

Index

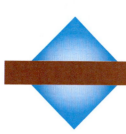

Index

A
absenteeism, 9.16-9.17
acetylene, 6.9
ACR. *See* air condition and refrigeration (ACR)
air as insulator, 4.2
air condition and refrigeration (ACR), copper tubing used in, 6.2, 6.4
alignment. *See also* misalignment
   pipe flanges to equipment nozzles, 3.16
   pipe to rotating equipment, 3.14-3.16
alloys
   melting points, 3.4
   soldering, 6.8
annealing, 3.13

B
balanced pressure trap, 4.3
ball joints, 5.2, 5.3, 5.16
base line, 2.17
benchmarks (BM), 1.11, 1.12
bimetallic trap, 4.3
blueprints, reading and interpreting
   following a single line, *bp3*, *bp9*, 1.16-1.30
   isometric drawings, *bp3*, *bp8*, 1.14-1.16
   piping and instrumentation drawings, 1.2-1.8, 1.12-1.14
   piping arrangement drawings, *bp2*, 1.8-1.12, 1.14
   site plans, *bp1*, 1.2
body, 8.2
bolts, flange, 8.5, 8.6
bonnet, 8.2
brazing copper tubing and fittings, 6.8, 6.10-6.13
brazing flux, 6.9, 6.11
bucket trap, inverted, 4.2, 4.8, 4.10, 5.6

C
Cam-Centric® valves, 8.2
capillary action, 6.7
carbon, 3.4
carbon dioxide in condensate, 4.2
carbonic acid, 4.2
chord, 2.25

circumference, segments in, 2.8
civil construction, 9.7
cold springing, 3.3-3.4
commercial construction, 9.7
Communication (Hutchinson), 9.4
communication styles, gender differences in, 9.4
compression couplings, 6.4
compression fittings, 6.2-6.4
compression joints
   hydraulic fitted, 6.24-6.29
   installing using copper tubing, 6.2-6.6
concrete, using in place of grout, 3.14
condensate, 4.2, 5.3
conductivity probes, 5.10-5.11, 5.16
constituents, 3.4
construction, categories of, 9.7
construction companies
   attracting qualified employees, 9.3-9.4
   chain of command, 9.2
   gender and minority issues, 9.4-9.7
   job descriptions, 9.8-9.9
   major departments in, 9.8
   policies and procedures, 9.8-9.9
copper fittings, soldering and brazing, 6.7-6.13
copper tubing
   hard vs. soft, 6.2
   installing flared and compression joints using, 6.2-6.6
   soldering and brazing, 6.7-6.13
couplings
   aligning, 3.14-3.15
   installing grooved pipe couplings, 6.33
craftworkers, 9.2
crayons, temperature indicating, 3.7, 3.10-3.12
crew leaders, 9.2
critical temperature, 3.5
cross line stops, 7.9
cryogenic plugs, 7.9
cultural conflicts in the workplace, 9.5
cutback, 2.8
cutback line layouts, 2.9-2.15, 2.22-2.23
cut grooving, 6.33

Note: For the purposes of this index, "*bp*" is a reference to the blueprint drawings that accompany the first module of Pipefitting Level Four.

D

decision making, supervisory, 9.13-9.17
desuperheaters, 5.14-5.16
detail sheet drawings, 1.24, 1.27
DeZurik® valves, 8.2
differential pressure, **5.7**
dimensions, 1.24
direct pressure, 8.2
disc, 8.2
discrimination, gender and minority, 9.6
drawing (tempering), 3.13-3.14
drip legs, 4.5, 5.3-5.5, 5.7
dummy legs and trunions, 2.29-2.32

E

elevation drawings, *bp6*, 1.9, 1.12
ell support, 2.37-2.43
employees
   attracting, 9.3
   problem, 9.16-9.17
   working together, 3.16
ethics, 9.12-9.13
expansion joints, 5.7, 5.8. *See also* thermal expansion
expansion loops, 3.2-3.3

F

Fabrications Materials list, 1.14
Federal Safe Drinking Water Act of 1986, 6.8
ferrule, 6.2, 6.4
filler metals, 6.11
filters, 5.7, 5.8, 5.16
fishmouth, 2.8, 2.23-2.29
fittings
   bends vs., 6.14, 6.15
   bolt-weld, 7.3-7.4
   bypass, 7.9
   compression, 6.2
   copper, soldering and brazing, 6.7-6.13
   equalization, 7.9
   flared and compression, installing, 6.4-6.6
   for hot tapping, 7.2-7.3, 7.9
   mechanical joint, 7.3
   split tee, 7.4-7.5, 7.9
   types of, 6.4, 6.5
flange bolts, 8.5, 8.6
flanges, flat-face to raised-face connection, 8.5
flared fittings, 6.4-6.6
flared joints, installing using copper tubing, 6.2-6.6
flare nuts, 6.3, 6.4
flaring, 6.3
flaring tools, 6.4, 6.5
flash steam, 4.3
float traps, 4.3, 4.8-4.9
flow
   measuring, 5.7-5.10
   stopping, 5.3
flow pressure switches, 5.11, 5.12, 5.16
flux, 6.8-6.9
freeze stops, 7.9
full port, 7.2-7.5

G

gaskets
   bonnet, 8.11-8.12
   selecting for grooved piping systems, 6.33
glass-lined piping, 6.21-6.24
grid line drawings, 1.11
grooved piping systems, 6.29-6.34
grout, basic types of, 3.14
grouting, primary reasons for, 3.14

H

handwheels, 8.8
hardenability, 3.7
head clearance, 7.2
heat affected zone (HAZ), 3.4, 3.5, 3.6
heaters and preheaters, 3.7-3.9
heating. *See also* preheating, stress relieving
   induction, 3.9
   interpass, 3.4
   tank coils, 2.7-2.8
heating equipment for soldering and brazing, 6.12-6.13
hole saws, 7.5
horizontal angularity misalignment, 3.15
horizontal offset misalignment, 3.15
horseshoe tool, 2.20-2.22
hot tap machines, 7.2, 7.5-7.7
hot taps
   installing fittings, 7.2-7.5
   line stop plugs, 7.8-7.10
   new technologies, 7.7
   safety and potential hazards, 7.2
hydraulic pipe bending, 6.20-6.21

I

Iconel® packing, 8.14
induction heating, 3.9
industrial construction, 9.7
in-line specialty equipment and devices
   ball joints, 5.2, 5.3, 5.16
   desuperheaters, 5.14-5.16
   drip legs, 5.3-5.5, 5.7
   expansion joints, 3.2-3.3, 5.7, 5.8
   filters, 5.7, 5.8, 5.16
   flowmeters, 5.7-5.10
   flow pressure switches, 5.11, 5.12, 5.16
   level measurement devices, 5.10-5.12
   protective equipment for installing or removing, 5.2
   rupture discs, 5.11, 5.13-5.14, 5.16
   safety devices, 5.11
   safety guidelines when installing or removing, 5.2
   snubbers, 5.2, 5.3
   storing and handling, 5.16
   thermowells, 4.9, 5.14
institutional construction, 9.7
interpass temperature control, 3.4, 3.10, 3.13
isometric drawings, 1.14-1.16, 1.30, 1.37

J

joint failure, causes of, 6.2
joints. *See also* expansion joints; flared joints
    brazing, 6.13
    flanged, 8.5
    leakproof, 6.9
    pressure-resistant, 6.11
    soldered, 6.7
    soldering, 6.9-6.10
joint strength, 6.8

L

language barriers in the workplace, 9.5
laterals, laying out, 2.25-2.29, 2.32-2.37
leadership
    ethics in, 9.12-9.13
    problem solving and decision making, 9.13-9.17
    styles of, 9.11-9.12
level measurement devices, 5.10-5.12
lines on blueprints, reading and interpreting
    following a single line, *bp3*, *bp9*, 1.16-1.30
    process flow and function lines, 1.5-1.6
    types of, 1.2
line stop plugs, 7.8-7.10
line stops, 7.2
liquid expansion trap, 4.3
Lokring® hydraulic compression system, 6.24-6.29

M

match lines, 1.6
metals
    critical temperature, 3.5
    expansion coefficients, 3.2
    hardening with quenching, 3.5
    melting point, 3.4
    nonferrous, 6.7, 6.8
    preheat temperatures, minimum, 3.5
    requiring preheating, 3.6-3.7
    temperature and structure of, 3.4-3.6
misalignment. *See also* alignment
    basic types of, 3.15
    results of, 3.2, 3.18
miter turn layouts
    calculating necessary elements, 2.15-2.16
    calculations for, 2.15-2.16
    four-piece, 90-degree, 2.18-2.20
    horseshoe tool for, 2.20-2.22
    three-piece, 90-degree, 2.17-2.18
    wyes, 2.22-2.23
miter turns, fabricating
    cutback line layouts in, 2.9-2.15, 2.22-2.23
    elements necessary for, 2.15
    four-piece, 90-degree, 2.18-2.20
    ordinate line layouts in, 2.8-2.10
    practice exercise, 2.15
    three-piece, 90-degree, 2.17-2.18
miter turns, layout calculations
    center-to-center length, 2.16
    degrees per cut, 2.15-2.16
    end piece length, 2.16
    number of segments, 2.15

N

nonferrous metal, 6.7, 6.8
normalizing, 3.13

O

offset misalignment, 3.15
offsets, determining
    right triangles in, 2.2-2.3
    simple offsets, 2.3-2.4
    tank heating coils, 2.7-2.8
    three line, 45-degree, equal-spread around a vessel, 2.4-2.5
    three line, 45-degree, unequal-spread, 2.5-2.7
open-top preheaters, 3.8
ordinate line layouts, 2.8-2.10
ordinate lines, 2.8
organizations, formal and informal, 9.7-9.8
orifice plates, 5.8-5.9, 5.16
orifice traps, 4.2
O-rings, 8.9-8.11
orthographic drawings, 1.14
oxidation, 6.8
oxyacetylene brazing equipment, 6.12-6.13

P

P&IDs. *See* piping and instrumentation drawings
packing, 8.2, 8.8, 8.13-8.14
pipe
    bending, 6.14-6.21
    copper, specifications, 6.3
    glass lined, 6.21-6.24
    plastic, hot tapping, 7.7
    undersea, hot tapping, 7.7-7.8
pipe bends, calculating and laying out, 6.16-6.19
pipe coupons, 7.5
pipe ends, preparing in grooved piping systems, 6.31-6.33
pipe fabrication
    fishmouth, 2.23-2.29
    laterals, laying out forty-five degree, 2.25-2.29, 2.32-2.37
    layouts without using references, 2.32-2.43
    miter turns, 2.8-2.23
    offsets, 2.2-2.8
    supports, 2.29-2.32, 2.37-2.43
piping, special
    bending, 6.14-6.21
    flared and compression joints, installing, 6.2-6.6
    glass-lined, 6.21-6.24
    grooved systems, 6.29-6.34
    hydraulic fitted compression joints, 6.24-6.29
    manufactured outside the U.S., 6.2
    safety and potential hazards when installing, 6.2
    soldering and brazing copper tubing and fittings, 6.7-6.13
piping and instrumentation drawings, reading and interpreting
    abbreviations on, 1.3-1.5
    general arrangement pages, 1.14
    introduction, *bp3*, 1.2, 1.12-1.13
    process piping, 1.2, 1.5-1.8

piping arrangement drawings, reading and interpreting
   control points, 1.9, 1.11
   coordinates, 1.9, 1.11-1.12
   elevations, *bp6*, 1.9, 1.12
   general arrangement pages, 1.14
   instrumentation symbology page, *bp5*, 1.9, 1.13-1.15
   mechanical symbology page, *bp4*, 1.13
   plan views, *bp7*, 1.8, 1.10
   sectional views, *bp2*, 1.8-1.9
   sectional view symbols, 1.11
piping systems
   determining offsets, 2.2-2.8
   flexibility, building in, 3.2, 3.3
   grooved, 6.29-6.34
   restraining points, 3.2
pitot tubes, 5.9, 5.10, 5.16
plug valves, 7.8-7.10, 8.2
port, 8.2
postheat, 3.4
postheat temperature control, 3.14
postheat weld treatment (PWHT), 3.4, 3.13-3.14
preheat, 3.4
preheating, stress relieving. *See also* heating
   equipment, 3.7-3.9
   metals requiring, 3.6-3.7
   minimum temperatures, 3.5
   preheating chart, 3.11-3.12
   preheat weld treatments, 3.4
   preventing hardening with, 3.5
   removing moisture through, 3.6
preheating devices, 3.7-3.8
preheat weld treatment (PHWT), 3.4
pressure
   direct, 8.2
   gauge, 4.8
   normal atmospheric, 4.8
   reverse, 8.2
pressure differential transmitters, 5.11, 5.12
pressure gauges, protecting, 5.2
pressure-relief valves, 5.13-5.14
problem solving, supervisory, 9.13-9.17
process piping drawings, reading and interpreting
   instrumentation, 1.8-1.9
   piping components and symbols, 1.6
   process equipment and symbols, 1.8
   process flow and function lines, 1.2, 1.5-1.6
project managers, 9.2
pyrometers, 3.9-3.10, 4.8, 4.9

Q

quality control, installing rupture discs, 5.11
quenching, 3.5, 3.13

R

reamer, 6.9
refrigerant piping, 6.8
residential construction, 9.7
reverse pressure, 8.2
revision, 1.24
revision marks, 1.24
rim-and-face alignment, 3.15
roll grooving, 6.31-6.33
rotameters, 5.10

rotating equipment, aligning pipe to, 3.14-3.16
rupture discs, 5.11, 5.13-5.14, 5.16

S

safety devices, 5.11, 5.13-5.14
safety policies, 9.17
seat, 8.2
sediment, collecting, 5.3
segment, 2.8
sexual harassment, 9.5-9.6
shop drawings, 1.14, 1.24
sight glasses, 5.11, 5.12
site plans, *bp1*, 1.2, 1.11
snubbers, 5.2, 5.3
solder, 6.2, 6.8-6.9
soldering copper tubing and fittings, 6.7-6.10
soldering flux, 6.8-6.9
sound as diagnostic technique, 4.8-4.10
sound measurement devices, 4.9
specials. *See* in-line specialty equipment and devices;
        piping, special
spool drawings, *bp3*, 1.14, 1.16, 1.24
steam
   desuperheaters, 5.14-5.16
   superheated, 4.3
   temperature measurement devices, 4.8, 4.9
steam header training system, 4.7
steam traps. *See also specific types of*
   condensate in, 4.2
   diagnosing and troubleshooting, 4.8-4.10
   failure, causes of, 4.10
   installing, 4.5-4.7, 5.5
   introduction, 4.0, 4.2
   maintaining, 4.10-4.11
   outdoor process, 4.7
   strainers, 4.3, 4.5, 4.6, 5.5, 5.7
   summary, 4.13
   types of, 4.2-4.3
steel
   hardenability, 3.7
   preheating requirements, 3.6-3.7
   quenching's effect on, 3.5
   welding temperature, 3.5
steel pipe, coefficient of expansion, 3.2
stems, 8.2, 8.8-8.9
STOPPLE® fittings, 7.8
stops
   cross line, 7.9-7.10
   low-pressure and vacuum, 7.9-7.10
strainers, 4.3, 4.5, 4.6, 5.5, 5.7
stress, 3.2
stress relieving
   for aligning pipe to rotating equipment, 3.14-3.16
   ball joints, 5.2
   expansion joints, 5.7
   interpass temperature control, 3.4, 3.10, 3.13
   for linear expansion, 5.7
   postheat temperature control, 3.14
   postheat weld treatments, 3.13-3.14
   preheating, 3.5-3.12
   temperature measurement devices, 3.7, 3.9-3.12
   for thermal expansion, 3.2-3.3
superintendents, 9.2

supervisors, roles and responsibilities
    basic skills, 9.2
    chain of command, 9.2
    decision making, 9.13-9.17
    delegation, 9.8
    leadership, 9.9-9.13
    problem solving, 9.13-9.17
    project workflow, 9.7
    safety, 9.17
    time allocation, 9.11
    training, 9.3-9.4
    understanding organization types, 9.7-9.10
switches, flow pressure, 5.11, 5.12, 5.16
symbols on blueprints, reading and interpreting
    instrumentation symbols, *bp5*, 1.8-1.9, 1.13-1.15
    mechanical symbology, *bp4*
    piping components symbols, 1.6
    sectional view symbols, 1.11

T

tanker vessels, hot-tapping sunken, 7.8
tank heating coils, 2.7-2.8
tapping valves, 7.3
T.D. Williamson, Inc., 7.8
Teflon®-encapsulated silicon O-rings, 8.10
Teflon® O-rings, 8.10
Teflon® packing, 8.13
temperature and metal structure, 3.4-3.6
temperature control, interpass, 3.4, 3.10, 3.13
temperature measurement devices
    pyrometers, 3.9-3.10, 4.8, 4.9
    temperature-indicating crayons, 3.7, 3.10-3.12
    thermocouple devices, 3.10
    thermowells, 4.9, 5.14
tempering, 3.13-3.14
thermal expansion, 3.2-3.4
thermocouple devices, 3.10
thermodynamic traps, 4.3, 4.5, 4.8-4.9, 4.11, 5.6
thermostatic traps, 4.3, 4.4, 4.8, 4.10-4.11, 5.6
thermowells, 4.9, 5.14
title block, 1.24
top hat, 5.5, 5.6
torches, 3.7-3.8
tracer, 4.3
traps. *See specific types of*
travel distance, 7.2
triangles, right, 2.2-2.3
trim, 8.2
trunions and dummy legs, 2.29-2.32
tubing
    copper, specifications, 6.3, 6.4
    types of, 6.2
turnover, employee, 9.16-9.17

V

Valmatic Cam-Centric® valves, 8.2
valve maintenance
    bonnet gaskets, 8.11-8.12
    introduction, 8.2
    packing and unpacking, 8.13-8.16
    stems and O-rings, 8.8-8.11
    summary, 8.18
    troubleshooting, 8.7-8.8

valves
    ball, 8.8
    diaphragm, 8.3
    flanged, 8.4-8.7
    gate, 8.8
    globe, 8.2
    importance of, 8.2
    line stop plugs, 7.8-7.10
    removing and installing, 8.2-8.7
    tapping, 7.3
    threaded, 8.3-8.4
valve stems, 8.2, 8.8-8.11
venturi tubes, 5.9-5.10
vertical angularity misalignment, 3.15
vertical offset misalignment, 3.15
vessel drawings, reading and interpreting, *bp8*, 1.16
vibration, grouting and, 3.14
vibration fatigue, 5.14
Viton® O-rings, 8.10

W

water hammer, 4.3
water supply piping, 6.8
welding. *See also* stress relieving
    hydrogen gas in the weld zone, 3.6
    instructions on spool drawings, 1.24
    quenching's effect on, 3.5
welding procedure specifications (WPS), 3.6
welds, dry washing, 3.14, 3.16
wetting, 6.8
witch's hat, 5.5, 5.6
women in construction, 9.4-9.6
wyes, 2.22-2.23